Harvard Studies in Technology and Society

The volumes in this series present the results of
studies conducted at the Harvard University Program
on Technology and Society. The Program was established
in 1964 by a grant from the International Business
Machines Corporation to undertake an inquiry in depth
into the effects of technological change on the economy,
on public policies, and on the character of the society,
as well as into the reciprocal effects of social progress on
the nature, dimensions, and directions of scientific and
technological developments.

Technological Change

Its Impact on Man and Society

EMMANUEL G. MESTHENE

Harvard University Press
CAMBRIDGE, MASSACHUSETTS

For Laura, Donna, and Jim

Preface

It is becoming commonplace that modern technologies—nuclear energy, rockets, computers, television, wonder drugs, and the latest surgical techniques—affect society in important ways. They serve to bring about changes in institutions and individual life styles; they generate strains for our values and beliefs; and they create problems—and opportunities—for our economic and political organizations.

New technologies have been bringing about social changes since the beginning of time, of course, and many of these were no doubt as fundamental and thoroughgoing for their times as anything we are experiencing today. The eighteenth-century Industrial Revolution, printing and gunpowder, the development of agriculture, and the invention of the wheel are examples that spring immediately to mind. What distinguishes our time in this respect, then, is less the fact that technology has important social consequences than our widespread awareness of that fact and our readiness to deal with it.

This new awareness is the result, partly of our improved understanding of the nature and consequences of modern science and technology, partly of our adeptness, as a society, in the deliberate use of technology to achieve our goals, and partly of sheer population growth. The last factor should not be underestimated. If contemporary American society is characterized by a trend toward bigness —in everything from the size of our cities to the size of the Federal budget—that is at least as much the consequence of how many of us there are as it is of our inventiveness in the development and use of sophisticated technologies. It is to technology *in conjunction with* population growth that we must look as we attempt to gauge the nature, dimension, and directions of contemporary social change, not to technology alone.

The widespread current concern with the implications of technology is leading scholars to turn their attention explicitly to an examination of the relationship between technological change and social change. A number of individuals are writing and offering university courses on the subject, and a few university centers are attempting to explore the area systematically, on a broad front, and in an interdisciplinary way.

Such explicit attention to the nature of the technology/society relationship is of recent origin, however. Few if any of the inquiries now underway date from before the 1960s. Add to that a subject matter that is wide-ranging, diffuse, and difficult to come to grips with, and it should surprise no one that we have learned little as yet compared to how much there is to learn. Much of what we have learned, moreover, is descriptive, or tailored to the relatively short-term policy needs of private management or public government. We have not made very much progress yet in explicating the mechanisms, that is, in tracing the cause-

effect relationships, by which technological innovation leads to changes in society. I am myself
convinced that such understanding of causes is a
necessary condition of *effective* knowledge about
technology and of effective policies for dealing with
its social consequences. To the extent that is true,
there is much work still to be done.

Yet we have been learning something in recent
years, even about the mechanisms of the tech-
nology/society interaction. The pages that follow
attempt to give a brief but systematic account of
what that something is. Inevitably, this particular
account relies heavily on the thinking and the
research projects that my colleagues and I are
engaged in at the Harvard University Program on
Technology and Society.[1] It is, however, couched
in a broader context that seeks to take account as
well of the work that others do.

The argument in brief is that new tools—it
takes some of the awe and mystery out of "technology" to think of it as tools—create opportunities
to achieve new goals or to do things in new ways.
This means that people and groups of people generally must organize themselves differently from
before in order to take advantage of the opportunities offered by new tools. (It takes some of the
awe and mystery out of the concept of "social
institutions" to think of them as groups of people
organized in certain ways to accomplish certain
purposes.)

Such reorganization, that is, such social change,

[1] All footnotes are grouped in a special Notes section beginning on
p. 91.

in turn means that older goals are often given up —or given less emphasis—in favor of new ones, and that certain values and beliefs are subjected to strain. This strain has two locations: It is to be found between old values and new techniques that somehow do not "fit" with each other; and it is also to be found between different groups of people, some of whom prefer the old values, goals, and techniques and others of whom are willing to give up the old because they find profit, or power, or preferred goals in the new techniques. It is just such strains that account for much of the social ferment that major technological changes usually occasion.

The strains that technology places on our values and beliefs, finally, are reflected in economic, political, and ideological conflict. That is, they raise questions about the proper goals of society and about the proper ways of pursuing those goals. In the end, therefore, the problems that technology poses (and the opportunities it offers) will be resolved (and realized) in the political arena, construing "political" broadly to include economic and ideological considerations as well as questions of more narrowly political organization and tactics. Technological innovation therefore leads ultimately to a need for social and political innovation if its benefits are to be fully realized and its negative effects kept to a minimum.

The effects of technology on society are thus much more complex, and demanding of much more knowledge and understanding, than is suggested by the currently fashionable—and contrasting—popular views about technology that are

described briefly at the opening of Chapter I below. Those views tend to be oversimplified and
biased, as is to be expected in an area about which
relatively little is known and in which the level of
discourse is not yet high. My hope in putting together what we and others are beginning to learn
about this area is to raise that level somewhat until further thought and study can raise it further
still.

I am indebted, for what I have learned from
them, to all with whom I have worked in the Harvard Program on Technology and Society. I am
grateful in particular to Harvey Brooks, Stanley
Dry, Paul R. Lawrence, Michael Maccoby, Robin
Marris, John R. Meyer, Richard S. Rosenbloom,
Joan Rothschild, and Alan F. Westin for reading
and commenting on manuscript drafts of this book.
My greatest debt is to Juergen Schmandt and Irene
Taviss, for helping me to write the better parts of
what follows but above all for joining me in the
rewarding enterprise of learning how technology
and society interact.

Contents

Technological Change

CHAPTER I

Social Change

Three Inadequate Views about Technology

Research aimed at discerning one or another of the particular effects of technological change on industry, government, or education is not new. Economists, social scientists, and other professional investigators have been engaged in it at least since Karl Marx, over a century ago. By contrast, systematic inquiry devoted to seeing all such effects together and to assessing their broader implications for contemporary society as a whole is relatively recent. Also, it calls for cooperation and joint effort among different academic disciplines, so that it lacks some of the methodological rigor and the richness of theory and data that mark the more established fields of scholarship.

As a result, efforts to understand and explicate the interaction between technological change and social change have to contend with a number of facile, one-dimensional or partial views about the nature of that interaction. It has always been so: Absence of knowledge encourages myth, or the comfortable illusion that there is nothing new to know.

One uncritical view that is prevalent at the present time holds that technology is a virtually unalloyed blessing for man and society. Technology is seen as the motor of all progress, as holding solutions for most of our social problems, as helping to liberate the individual from the clutches of a complex and highly organized society, and as the source of permanent prosperity—in short, as the promise of utopia in our time.

This view has its modern origins in the Baconian conception of knowledge as power, in the social philosophies of such nineteenth-century thinkers as Saint-Simon and Auguste Comte, and probably also, at least for Americans, in the pragmatic conviction that there is nothing a nation of doers cannot do. The view tends nowadays to be held by many scientists and engineers, by many military leaders and aerospace industrialists, by people who believe that man is fully in command of his tools and his destiny, by people who think he should be if he is not, and by many of the devotees of such modern techniques of "scientific management" as systems analysis and program planning and budgeting.

On its surface, this view exhibits the optimism that we associate with the rationalistic tradition in western intellectual history, as in the eighteenth-century Enlightenment in France, for example. It places great faith in the social efficacy of scientific methods and tools, and it by and large assumes a model of society according to which science is the principal determining element in the shape and destinies of men and their institutions.

Below the surface, one may detect traces of economic and political ideology in this view. Some vested interests find profit or other advantage in new technology, and many who would bend society to their purposes often see it as a means of bringing about deliberate social change.

A contrary view sees technology as an almost unmitigated curse. Technology is said to rob people of their jobs, their privacy, their participation in democratic government, and even, in the end, of their dignity as human beings. It is seen as autonomous and uncontrollable, as fostering materialistic values and as destructive of religion, as bringing about a technocratic society and bureaucratic state in which the individual is increasingly submerged, and as threatening, ultimately, to poison nature and blow up the world.

This view is heir to two different traditions. It is akin to historical "back-to-nature" attitudes toward the world, such as we associate with Jean-Jacques Rousseau and Henry Thoreau. It also derives from traditional socialist critiques of the appropriation of technology as capital. From one or another of these standpoints, this view is currently propounded by many artists and literary commentators, by popular social critics, and by existentialist philosophers and latter-day Marxists. It is attractive to many of our youth, and it tends to be held, understandably enough, by segments of the population that suffer dislocation as a result of technological change.

By contrast with the optimistic view of technology, this one is marked by the pessimism that has often been associated with the mystical tradition in the history of the West. Taken together, the two views may be seen as the latest stage of the eternal battle between God and the Devil. In the first view, technology is invested with an omnipotence heretofore reserved to the Almighty. In the second view, technology emerges as the modern counterpart of the Devil, responsible, as the Devil has traditionally been, for "man's eternal inhumanity to man." The psychological counterpart of this contrast, as the two views make clear, is that between optimists and pessimists. In sociological-political-economic terms, we are dealing with the contrast between those who have power and status and those who do not.

There is a third view, which is of a quite different sort, however, and the inadequacies of which derive from its being partial and hypercritical, rather than biased and uncritical. This view holds that technology as such is not worthy of special notice. The arguments advanced for this conclusion include the following: Technology is not new, and it has moreover been recognized as a factor in social change at least since the Industrial Revolution; it is unlikely that the social effects of modern technology—even of computers, for example—will be nearly so traumatic as the introduction of the factory system into eighteenth-century England; research has shown that technology has done little to accelerate the rate of economic productivity since the 1880s; and, there has been no significant change in recent decades in the time period between invention and widespread adoption of new technology.

Moreover, according to this argument, improved communications and higher levels of education make people today much more adaptable than ever before to new ideas and to the new social reforms required by technology. If anything, therefore, technological change is likely to be less upsetting than in the past, because its scope and rate are roughly in equilibrium with man's social and psychological development.

Although this view is supported by a good deal of empirical evidence, it tends to underemphasize a number of the characteristics of modern technology and of the modern world that deprive historical comparison of some of its force. Among these are the effects of sheer physical power, speed of communication, and population densities, which are less easy to identify and measure with precision than, say, more strictly economic changes. Such underemphasis of relevant factors reflects the difficulty of coming to grips with a new and broadened subject matter by means of concepts, methods, and intellectual categories that were designed to deal with older and different subject matters.

This third view tends to be held by historians, for whom continuity is an indispensable methodological assumption, and by many economists, who find that their techniques measure some things quite well while those of the other social sciences are as yet much less refined and reliable.

Stripped of the trace of caricature I have allowed to creep into my descriptions, each of the three views I have discussed contains a measure of truth and evokes a genuine aspect of the relationship of technology and society. Yet they are oversimplifications that do not really serve to advance understanding. One can find empirical evidence to support each of them—and especially so the third—without however gaining much knowledge about the mechanism by which technology leads to social change, or much insight about the implications of technology for the future. The first two views remain too uncritical, and the third too partial, to be reliable guides for inquiry. Recent research and analysis have been leading to more differentiated conclusions and revealing more subtle relationships, as I hope what follows will show.

Some Countervailing Considerations

It is important to recall that technology is not independent of the society in which it develops and flourishes. To fail to do so—as both the overoptimistic and overpessimistic views described above tend to do—can be to see technology as a somehow autonomous force that develops according to its own internal laws and lets its consequences fall where they may. The third view is sounder in this respect. It displays a more explicit awareness of the social origins of even the most sophisticated technologies, and it recognizes the role of social institutions both in fostering technological development and in mediating between technology and its effects.

The optimistic view of technology, for example, emphasizes the "revolutionary" or radically beneficial possibilities inherent in new technology. There is no doubt that the possibilities are there, but it does not follow that they are easily realized just because they are there. The social setting into which technology is introduced has much to do with whether, to what degree, in what ways, and which ones of the benefits potential in it will be achieved.

This has been well illustrated in a recent study of computer-based educational technology by Anthony G. Oettinger.[2] The author, who is an eminent computer scientist, recognizes both the exciting possibilities and the long-term promise of experiments with the use of computers, assorted teaching machines, and systems analysis in secondary education. When he turns from experimental possibility to large-scale practical application, however, he concludes that while education is badly in need of reform the probability is slight that its ills will soon yield to the kind of quick technological fix advocated by the most enthusiastic proponents of educational innovation.

Professor Oettinger examines both the latest educational hardware and the educational institutions that would take advantage of it. Both are found wanting. The hardware itself is as yet much more primitive than is generally appreciated, so that fragile, unreliable, and expensive devices often gather dust in a classroom corner once the enthusiasm that greeted their arrival has subsided. Knowledge about how to apply the technology is even more primitive; teaching methods and curriculum contents remain virtually unmodified by the availability of new devices.

The biggest obstacle to the rapid and effective introduction of technology into the schools, however, is the structure of the American school system itself, which, in Oettinger's words, "seems ideally designed to resist change." That system succeeds in combining the rigidity of a military service and the fragmentation of small business, without either the centralized authority that can ultimately make the military move, or the initiative and flexibility of response enjoyed by the innovative entrepreneur.

The study concludes that neither educational technology nor the school establishment is ready to consummate the revolution in learning that will bring individualized instruction to every child, systematic planning and uniform standards to 25,000 separate school districts, an answer to bad teachers and unmovable bureaucracies, and implementation of a national policy to educate every American to his full potential for a useful and satisfying life. Technology may be the motor of all progress, but institutional sluggishness will most often turn out to be a very effective brake.

The brake, fortunately, works both ways. The negative possibilities that the pessimistic critics find to be inherent in technology must also be mediated by social institutions and are therefore not inevitable. This conclusion is suggested by a study currently being conducted by Professor Manfred Stanley of Syracuse University.[3]

Professor Stanley is examining the nature and roots of the pessimistic literature about technology, as exemplified by such writers as Jacques Ellul,

Hannah Arendt, Lewis Mumford, and Herbert Marcuse. These writers are widely read, partly, no doubt, because they touch on the fears and unanswered questions of many people. They predict the emergence of social planning by a centralized state bureaucracy and an ultimate fusion of state and society under domination of a technocratic elite. These trends are seen as leading inevitably to the kind of world described in the disutopias of Aldous Huxley and George Orwell, in which the individual human being is reduced to a mere cog in a social machine whose course is defined by the logic of the technical process itself.

Professor Stanley points out, however, that other writers have come to opposite conclusions based on the same trends. The Marxists and Karl Mannheim, for example, regarded social planning and an effective administrative state as the only possible basis for genuine democratic pluralism. Others see the development of government by experts as the only way in which a democracy can deal effectively with highly technical policy issues, provided only that mechanisms of consultation and accountability are kept effective. Still others, like Thomas Luckmann and Paul and Percival Goodman, see possibilities of enhanced individual self-determination or of strengthened human community in some of the developments we associate with a technological society.

There is no doubt that there are real destructive possibilities resident in technological change, as anyone knows who has breathed smog, bathed in polluted water, or contributed taxes to the international arms race. But the argument that converts this fact into a prediction of inevitable doom contains large doses of ideological rhetoric and reveals a number of preexisting value commitments on the part of those who make it.

Professor Stanley's study is leading to the conclusion that such predictions are premature, since there are alternative social outcomes potential in the process of technological change. In other words, the range of possibility and of human choice implicit in technology is greater than the pessimistic critics assume. The problem—here, as well as in the application of educational technology—is how to organize society so as to free the possibility of choice and how to control our technology wisely in order to minimize its negative consequences.

The third view that I discussed above amounts to questioning whether modern technology and its effects constitute a distinct subject matter deserving of special attention. The answer depends largely on how technology is defined. At the Harvard Program on Technology and Society, we construe the term as including more than hardware alone. Most serious investigators have concluded that understanding is not advanced by concentrating on such apparently meaningful yet imprecise questions as "What are the social implications of computers, or lasers, or space technology?" Society and the influences of technology upon it are much too complex for such artificially limited approaches to be useful. The opposite error made by some people is to define technology too broadly by identifying it with rationality in the broadest sense. The term is then operationally meaningless and unable to support fruitful inquiry.

We have found it more useful to define technology as tools in a general sense, including machines, but also including such intellectual tools as computer languages and contemporary analytic and mathematical techniques. That is, we define technology as the organization of knowledge for the achievement of practical purposes. It is in this broader meaning that we can best see the extent and variety of the effects of technology on our institutions and values. Its pervasive influence on our very lives and culture would be unintelligible if technology were understood as no more than machines.

It is in the pervasive influence of technology that our contemporary situation seems qualitatively different from that of past societies. Therein lies the counterargument to the third view of technology. One reason for this difference is that our tools are more powerful than any before. The rifle wiped out the buffalo, but nuclear weapons can wipe out man. Dust storms lay whole regions waste, but too much radioactivity in the atmosphere could make our planet uninhabitable. The domestication of animals and the invention of the wheel literally lifted the burden from man's back, but computers could free him from all drudgery.

This quality of finality of modern technology, and the degree to which our time is oriented toward and dependent on science and knowledge, have brought our society, more than any before, to explicit awareness of technology as an important determinant of our lives and institutions. As a result, our society is coming to a deliberate decision to understand and control technology, and is therefore devoting significant effort to the search for ways to measure the full range of its effects. It is this prominence of technology in many areas of modern life that seems novel in our time and deserving of explicit attention.

How Technological Change Impinges on Society

It is clearly possible to sketch a more adequate hypothesis about the interaction of technology and society than any of those implicit in the three inadequate views discussed above. Such a hypothesis would begin with the observation that technological change induces or "motors" social change in two principal, closely interrelated ways. New technology creates new opportunities for men and societies, and it also generates new problems for them. It has both positive and negative effects, and it usually has the two *at the same time and in virtue of each other.*

Thus, for example, industrial technology—that is, new machines and processes and the advent of factory automation—strengthens the economy, as our measures of growth, productivity, and living standards all show. A recent study of structural changes in the American economy also shows, however, that the introduction of new industrial technology changes the relative importance of individual supplying sectors in the economy.[4] That is, new machines and techniques of production alter the amounts and kinds of materials, of parts and components, of energy, of labor skills, and of supporting services that each industry uses to manufacture its products. Instances of this kind of change in recent years have been the shift from coal to oil and natural gas for residential heating and the marked displacement of steel and tin by paper and plastics in the container industry.

All such changes bring about dislocations of businesses and of employment patterns, which are the negative counterparts of technological development and economic growth. It is within this framework, incidentally, that the effects of technology on work and employment are best understood. The critics of technology are fond of saying that automation deprives people of work and may soon reduce the majority of our population to idleness. There is neither evidence nor theoretical warrant for this proposition. As the 1966 report of the President's Automation Commission pointed out, "the general level of unemployment must be distinguished from the displacement of particular workers at particular times and places." [5]

For example, jobs for makers of wagon wheels and horseshoes were virtually eliminated when the technology of the automobile took hold. But the total number of jobs in the nation was not reduced thereby. Many more workers are engaged in automobile manufacture, in fact, than were ever involved in equipping horses and carriages. A new machine or product, or a new automatic process, can make some jobs or skills obsolete, but others are created as new industries develop and new products are marketed. The transitional problems that technology creates cannot be separated from the new opportunities that it also creates.

The actual process by which technological change brings about social change helps to explain why any given technological development is likely to combine positive and negative effects. The process usually consists of a number of events occurring in sequence. A new invention or technological development—a new tool, in short—generally creates a new opportunity, either to do something differently or better than before, or to do something for the first time that simply was not possible at all before. Some previously distant aspiration may thus be converted into an achievable goal.

Inevitably, when that happens, there will be people, or groups of people (that is, organizations or institutions), or whole societies that will be motivated to capitalize on the new opportunities. The motivation may be purely selfish, of course, as in the desire to turn the new technology to profit. But it may also be entirely altruistic, as in realization of the goal of adequate nutrition for the world's population that may be potential in nuclear-powered desalinization technology.

Whatever the motivation, realization of the opportunities inherent in the new technology (unless it represents only a small marginal improvement over previous technologies) will call for changes in social organizations, that is, in the ways in which people and institutions are organized to accomplish their purposes. (This will surprise no one who has had the experience of learning the way to handle a new tool previously unavailable to him.) Such new organizations do not spring from nothing, however. They grow up alongside, or they replace, previously existing social structures, which are themselves organized to use earlier technologies in the achievement of previously defined goals. To the extent that the new organizations compete effectively with the older ones for economic resources and political status, the goals served by the older structures may be less adequately achieved than they once were. The gain realized by achievement of new goals, in other words, involves some loss in the realization of older goals, and it is the *same* technological/social process that brings about both the gain and the loss.

A study of regional and urban economics by John R. Meyer and John F. Kain illustrates this process by tracing some of the consequences of changes in transportation technology.[6] The opportunities for increased mobility of businesses and individuals that are inherent in automobile technology, for example, are easy to see and appreciate. Exploitation of those opportunities in the form of increased ownership of private automobiles has however led to altered patterns of industrial and residential location in our urban centers, so that older unified cities are being increasingly transformed into larger metropolitan complexes. The new opportunities for mobility are relatively less available to the poor and black populations of the core cities than to white residents, however, partly for economic reasons and partly as a result of restrictions on choice of residence by black families.

This results in persistent unemployment among blacks, despite a generally high level of economic activity in the nation as a whole, with the further consequence that cities are increasingly unable to perform their traditional functions of providing employment opportunities for all segments of their populations and an integrated social environment that can temper ethnic and racial differences. The new urban complexes are neither fully viable economic units nor effective political organizations able to upgrade and integrate their core populations into new economic and social structures. The resulting instability is further aggravated by modern mass communication technology, which heightens the expectations of the poor and the fears of the well-to-do and adds frustration and bitterness to the urban crisis.

An almost classic example of the sequence of steps by which technology impinges on society has been provided by a study of changes in the system and practice of medical care, conducted by Mark G. Field of Boston University.[7] Recent advances in biomedical science and technology have created two new opportunities: They make possible treatments and cures that were never possible before; and they create conditions under which adequate medical care can be provided to the population at large as a matter of right rather than privilege. Such changed conditions are often the result of the economic affluence that comes with changes in industrial technology.

In realization of the first possibility, the medical profession has become increasingly differentiated and specialized and is tending to concentrate its best efforts in a few major urban centers of medical excellence. This alters the older social organization of medicine that was built around the general practitioner. The second possibility has led to big increases in demand for medical services, partly because a healthy population has important economic advantages in a highly industrialized society. This increased demand accelerates the process of differentiation and multiplies the levels of paramedical personnel between the physician at the top and the patient at the bottom of the hospital pyramid.

Both of these changes in the medical system are responsive to the new opportunities for technical excellence that have been created by biomedical technology. Both also involve a number of well-known costs, however, in terms of some older desiderata of medical care. The increasing scarcity of the general practitioner in many sections of the country means that people in need often have neither easy access to professional care nor the advantage of a "medical general manager" to direct them to the right care at the right place at the right time. This can result both in poor treatment and a waste of medical resources.

Further, too exclusive an emphasis on technical excellence can lead to neglect of the patient's psychological well-being, partly because the physician finds the technical cure more certain and immediate and partly because of the degree of "indignity" implied by increasing dependence on mechanical and electronic devices for the maintenance of human life and personality. Development of artificial organs may thus in itself turn into a micro-example of a technology that brings with it both positive physical and negative psychological effects at the same time.

One may note in passing that further technological development can often carry the solution to problems created by an earlier technology. In medicine specifically, there is the possibility that full and wise use of computer technology can at some time help to redress some of the imbalances noted in Professor Field's study. In the best of all possible worlds, the physician might be rendered free to devote his principal attention to the "whole" patient, while computers and associated technologies carry the purely technical burden of diagnosis and routine treatment. Such a new technological development would in turn call for still further social changes in the organization of medicine, in the selection and training of physicians, and even, perhaps, in the physical system for delivery of medical care. It is not inconceivable that computer centers and closed circuit television networks might once again reduce the importance of large hospitals relatively to decentralized care in the home or in small community health centers.

The pattern illustrated by the preceding examples tends to be the general one. It is clearly evident in our most spectacular technological successes in America in the last quarter of a century, that is, in national defense, in space exploration, and in the provision of consumer goods and services. These successes have provided protection for the nation, realization of an age-old human dream, and achievement of the highest standard of living ever enjoyed by man. They have also brought with them, however, enthusiastic advocates and vested interests who claim that development of sophisticated technologies and technological products is an intrinsic good that should be pursued for its own sake, and they have perpetuated a system of economic organization that is insufficiently motivated and ill-equipped to anticipate many of the destructive consequences of modern technology. It is this negative side of our positive technological achievements that contributes to the self-reinforcing and antisocial character that many people associate with technology, and that raises fears of an autonomous Frankenstein's monster uncontrollable by man.

Mass communication technology has also made rapid strides since World War II, with great benefits for education, journalism, commerce, and sheer convenience of living. It has also been responsible for a good deal of social and international unrest, however, by raising expectations faster than our political forms can deal with them. It may moreover help explain the singular rebelliousness of a youth that can find out what the world is really like from television before home and school have the time to instill some ethical sense of what it should or could be like.

That is, departures from the ideals and norms of society are displayed by the mass media before a youngster can learn what values and value accommodations the ideals and norms reflect, and before he has the knowledge of history, contemporary fact, and future possibilities to support sound judgment and meaningful action. This may account in large part for the typically anti-intellectual and anarchistic character of the radicalism of the sixties, by comparison with the highly doctrinaire and utopian Marxist radicalism of the years before World War II.

These various examples illustrate what was noted above, that technological change creates new opportunities and new problems at the same time and in virtue of each other. That is why isolating the opportunities (as the technological optimists do) or the problems (as the pessimistic critics are prone to) and construing one or the other as the whole answer is ultimately obstructive of rather than helpful to understanding.

On Taking Advantage of Technological Opportunities

The heightened prominence of technology in our society faces us with the interrelated tasks of profiting from its opportunities and containing its dangers. At the present time, much public and political concern is being devoted to the increasingly visible and distressing negative impacts of modern technology, on the environment as well as on man. The concern is deserved, but it is important to give at least equal attention to deriving the benefits for society that are also potential in technology.

This is not to suggest that we have not done so in many cases, of course: in space and in industrial production, as already noted, as well as in such areas as medicine and communications. In all such cases, we have been able to achieve a successful combination of technical inventiveness, adequate resources, and institutional adaptiveness. All three of those ingredients must be present. In areas in which they are not, there is a presumption that society has failed to take full advantage of its technological opportunities.

Failure of society to respond to the opportunities created by technological change means that much actual or potential technology lies fallow, that is, is not used at all or is not used to its full capacity. This can often mean that potentially solvable problems are left unsolved and potentially achievable goals unachieved, because we waste our technological resources or use them inefficiently.

The best hope of controlling the damaging effluent of the old technology of internal combustion, for example, lies with newer electric, chemical, or filtration technologies, and provision of adequate low- and medium-income housing could be moved off dead center if we would develop the political means of taking advantage of new materials and construction technologies. A society has at least as much stake in the efficient utilization of technology as in that of its natural or human resources.

There are often good reasons, of course, for not developing or utilizing a particular technology. The mere fact that it can be developed is not sufficient reason for doing so. The costs of development may be too high in the light of the expected benefits, as was the case a few years ago with the project to develop a nuclear-powered aircraft. Or, a new technological device may be so dangerous in itself or so inimical to other purposes that it is never developed, as in the cases of Herman Kahn's "Doomsday Machine" and the recent proposal to "nightlight" Vietnam by reflected sunlight.

But there are also cases where technology lies

fallow because existing social structures are inadequate, and prevailing value systems poorly at-
tuned, to exploiting the opportunities it offers. A
clear example of this lack is provided in a recent
examination of institutional failure in the urban
ghetto by Richard S. Rosenbloom and Robin Marris.[8] At point after point this study confirms what
has long been suspected, that is, that traditional
institutions, attitudes, and approaches are by and
large incapable of coming to grips with the new
problems of our cities. Many of those problems
were themselves caused by technological change,
as was shown by Meyer and Kain in the study
previously referred to, but existing social mechanisms seem unable to realize the possibilities for re-
solving them that are also inherent in technology.

Vested economic and political interests serve to
obstruct adequate provision of low-cost housing.
Community institutions wither for want of interest
and participation by residents. City agencies are
unable to marshal the skills and take the systematic
approach needed to deal with new and intensified
problems of education, crime control, and public
welfare. Business corporations finally, which are
organized around the expectation of private profit,
are insufficiently motivated to bring new technology
and management know-how to bear on urban
projects where the benefits appear to be largely
social. Business has yet to estimate the costs of
racial discrimination in terms of decreased pur-
chasing power and its consequent depressing effect
on the private economy.

All these factors combine to dilute what could
otherwise be translated into a genuine public desire

to apply our best knowledge, our latest tools, and adequate resources to the resolution of urban tensions and the eradication of poverty in the nation.

There is also institutional failure of another sort. Government in general and agencies of public information in particular are not yet equipped for the massive task of public education that is needed if our society is to make full use of its technological potential, although the Federal government has been making significant strides in this direction in recent years. Thus, much potentially valuable technology goes unused, because the public at large is insufficiently informed about the possibilities and their costs to provide support for appropriate political action. To use a negative example, it was an informed and therefore aroused public opinion that finally led to governmental safety regulation of the automobile industry. More positively, it is education of the public in the potentialities of construction technology and in the barriers to its full utilization embodied in zoning ordinances and in the interests of the affected industries and labor unions that seems necessary for adequate formulation and implementation of urban housing programs.

The general point is that political action in a democracy depends on the availability of abundant, accurate, and widely disseminated political information. As noted, we have done very well with our technology in the face of what were or were believed to be crisis situations, as with our military technology in World War II and with our space efforts when beating the Russians to the moon was deemed a national goal of first priority. We have also done very well when the potential benefits of technology were close to home or easy to see, as in improved health care and better and more varied consumer goods and services.

We have done much less well in developing and applying technology where the need or opportunity has seemed neither so clearly critical nor so clearly personal as to motivate political action, as in the instance of urban policy discussed above, or in that of educational technology for that matter. Where technological possibility continues to lie fallow, it is to improved political information and to innovation in political institutions that society must attend if it is to use its tools to their full effectiveness.

Containing the Negative Effects of Technology

The kinds and magnitude of the negative effects of technology are no more independent of the institutional structures and cultural attitudes of society than is realization of the new opportunities that technology offers. With few exceptions, technologies in our society are developed and applied as a result of individual decision. Individual entrepreneurs, individual firms, and individual government agencies are always on the lookout for technological opportunities: either new machines to reduce the costs of production, or new mechanical products to sell, or new technological systems to facilitate accomplishment of a mission. Technological innovation is so much sought after, in fact, that the Federal government puts $16 billion a year into research and development and big companies hire whole staffs of scientists and engineers to create new technologies.

In deciding whether to develop a new technology, the individual decision maker calculates the benefit that he can expect to derive from the new development and compares it with what it is likely to cost him. If the expected benefit to himself is greater than the cost he will have to pay, he goes ahead. In making this calculation, the individual decision maker does not pay very much attention to the probable benefits and costs of the development to others than himself or to society generally, except to the extent that he is required to do so indirectly by relevant laws or governmental regulations. These latter benefits and costs are characterized by economists as *external*.

The external benefits potential in new technology —what I referred to in the preceding section as technological opportunities—may thus not be fully realized by the individual developer. They will rather accrue to society as a result of deliberate social action, as pointed out above.

Similarly with the external costs. The direct costs of technological development include design and engineering, labor, raw materials, plant and equipment, packaging, marketing, distribution, and so forth. The individual developer pays for these. But he does not pay—or he pays very little and very indirectly—for the waste that he pours into the river, the smoke he discharges into the air, the job dislocations that he causes when he automates his plant, or the noise nuisance that may go with a new airplane design. In minimizing only expected costs to himself, in other words, the individual decision maker will tend to contain only some of the potentially negative effects of the new technology. The external costs, and therefore the negative effects on society at large, are not of principal concern to him, and in our society they are generally not expected to be.

Most of the consequences of technology that are causing concern at the present time—the proliferation of weapons technology, smog, water pollution, radioactivity, urban sprawl, sonic booms, threats to the beauty and balance of nature, social and psychological tensions and unrest, job dislocations, and encroachments on individual privacy—are negative externalities of this kind. They are with us in large measure because it has not been anybody's explicit business to foresee and anticipate them. They have fallen between the stools of innumerable individual decisions to develop individual technologies for individual purposes without anyone— any organization or agency—around to give explicit attention to what all these decisions add up to for society as a whole and for people as human beings.

This freedom of individual decision making is a value that we have cherished and that is built into the institutional fabric of our society. The negative effects of technology that we deplore are a measure of what this traditional freedom is beginning to cost us. They are traceable less to some mystical autonomy presumed to lie in technology and much more to the autonomy that our economic and political institutions grant to individual decision-making. Technology would seem much less autonomous if we could internalize all its costs and charge them directly to the producer. They would then be paid for either by the ultimate consumer (in the form of higher prices in the case of consumer goods or of some sort of fee, as in a toll-road) or by society as a whole when necessary (in one form or another of public subsidy), but only following a deliberate and explicit public decision to do so.

When the social costs of virtually unrestricted individual decision making in the economic realm achieved crisis proportions in the great depression of the 1930s, the Federal government introduced economic policies and measures many of which had the effect of abridging the freedom of individual decision. Now that some of the negative impacts of technology are threatening to become critical, the government is considering measures of control that will have the analogous effect of constraining the freedom of individual decision makers to develop and apply new technologies irrespective of social consequence.

The U.S. House of Representatives, for example, is considering a proposal to establish a National Technology Assessment Board to help identify and anticipate the expected effects of proposed technological developments and assist in evaluating their probable social utility (or disutility).[9] As one observer has noted, "An assumption underlying this orientation is that man can exercise an appreciable degree of control over social development, [and] that technology does not have to be utilized simply because it exists and without regard to [its] effects on the total social environment."[10]

In the U.S. Senate, there is a proposal to "establish a select Senate committee on technology and the human environment" to study and assess the impact of technology "on man's thinking, health, work, living habits, [and] individuality," and to help provide "a base for the development of national goals for environmental betterment of man in America and the rest of the world." [11]

In the executive branch, finally, attention is being directed to development of a system of social indicators to help gauge the social effects of technology, to establishment of some body of social advisers to the President to help develop policies in anticipation of such effects, and generally to strengthening the role of the social sciences in public policy-making.

The effort to develop a system of social indicators is responsive to the need cited above for a way of measuring the full range of the social effects of technology, rather than only those bearing on the economy. The annual Economic Report of the Council of Economic Advisers is replete with indicators of economic growth, productivity and employment, inflation, expenditures, investment, consumption, and income distribution. There are no similar indicators of the national health, of employment opportunities, of the state of the environment, of the degree and location of poverty, of the costs and effects of crime, of the richness of our education, our science, and our art, or of the degree and effects of participation in or alienation from public life.

The seven indices listed immediately above are taken from the table of contents of a recently issued government report entitled *Toward a Social Report*.[12] One of the principal authors of this report speaks of it as follows:

No society in history has as yet made a coherent and unified effort to assess those factors that, for instance, help or hinder the individual citizen to establish a career commensurate with his abilities, or to live a full and healthy life equal to his biological potential, which define the levels of an adequate standard of living, and which suggest what a "decent" physical and social environment ought to include. The document *Toward a Social Report* is the first step in the effort to make that assessment.[13]

Toward a Social Report does not deal with the measurement and assessment of the specific impacts of technological change, but there is no question that it provides a sounder basis for doing so in the future than any that has existed thus far.

All such attempts to assess, control, and mitigate the negative effects of technology appear to many to threaten freedoms that our traditions still take for granted as inalienable rights of men and good societies, however much such freedoms may have been tempered in practice by the social pressures of modern times. These include the freedom of the market, the freedom of private enterprise, the freedom of the scientist to follow the truth wherever it may lead, and the freedom of the individual to pursue his fortune and decide his fate.

There is thus set up a tension between the need

to control technology and our wish to preserve our values, which leads some people to conclude that technology is inherently inimical to human values. The political effect of this tension takes the form of inability to adjust our decision making structures to the realities of technology so as to take maximum advantage of the opportunities it offers, and so that we can act to contain its potential ill effects before they become so pervasive and urgent as to seem uncontrollable.

To understand why such tensions are so prominent a social consequence of technological change, it is necessary to look explicitly at the effects of technology on social and individual values.

CHAPTER II

Values

The Challenge of Technology to Values

Despite the great importance of scientific techniques and of the processes and institutions of knowledge in contemporary society, it is clear that political decision making and the resolution of social problems are not dependent on knowledge and rational method alone. Many commentators have noted that ours is a "knowledge" society that is devoted to an "end of ideology" and that bases its decisions on the collection and analysis of data, but none would deny the role that values play in shaping the course of society and the decisions of individuals.[14]

On the contrary, questions of values can become more pointed and insistent in a society that organizes itself to control technology and that engages in deliberate social planning. Planning demands explicit recognition of value commitments and value priorities and often brings into the open value conflicts that remain hidden in the more impersonal and less conscious workings of the market system.

45

In economic planning, for example, we have to make explicit choices between the value we attach to leisure and the value we place on increased economic productivity, and we have to do so moreover without having a common measure on the basis of which the choice can be made. In planning education, we come face to face with the traditional American value dilemma of equality versus achievement: Do we opt for equality and nondiscrimination by giving all students the same basic education, or do we emphasize freedom and foster achievement by tailoring education to individual learning capacities, which are themselves often conditioned by social and economic background?

Current science-based decision making techniques also call for clarity in the specification of goals, thus serving to make value preferences explicit. The effectiveness of systems analysis, for example, requires either that objectives and criteria of evaluation be known in advance, or that alternative possible objectives be clearly enough formulated so that they can be compared; and criteria and objectives of specific actions obviously relate to a society's system of values. That, incidentally, is why the application of systems analysis meets with less relative success in educational or urban planning than in many facets of military planning.

The increased awareness of conflicts among our values that planning and rational decision making produce serves in part to explain the generally questioning attitude toward traditional values that appears to be endemic to a high-technology, knowledge-based society. Another part of the explanation lies in the discovery that values once

considered immutable are often based on inadequate knowledge. As we learn more about the
physiological and psychological bases of deviant behavior, for example, we begin to question ancient
concepts of personal responsibility and the value of
punishment as a deterent to crime. In general, as
Robin Williams has noted,

> A society in which the store of knowledge
> concerning the consequences of action is
> large and is rapidly increasing is a society
> in which received norms and their "justi-
> fying" values will be increasingly subjected
> to questioning and reformulation.[15]

This is another way of pointing to the tension,
alluded to at the end of the last chapter, between
the need for social action based on knowledge on
the one hand and the pull of our traditional values
on the other. The increased questioning and reformulation of values that Williams speaks of,
coupled with a growing awareness that our values
are in fact changing under the impact of technological change, leads many people to believe that technology is by nature destructive of values. But this
belief presupposes a conception of values as eternal
and tends to confuse what is valuable with what is
unchanging.

To be sure, the values of a society change more
slowly than do the realities of human experience;
their persistence is inherent in their nature as values
and in their function as criteria of judgment and
action, which means that their own adequacy will
tend to be judged later rather than earlier. But

values do change, as a glance at history shows. Some are abandoned as circumstances change and new ones are formulated to deal with new situations. Most frequently, we make rearrangements in our value hierarchy; values once considered crucial become less relevant and therefore less important, while others, once relatively lower in our estimation, take on new importance. Values do not have to be eternal and unchanging in order to be values.

The fact that values come into question as our knowledge increases, therefore, and that some traditional values cease to function adequately when technology leads to changes in social conditions does not mean that values as such are being destroyed by knowledge and technology. What does happen is that values change through a process of accommodation between the system of existing values and the technological and social changes that impinge on it. Understanding of the effects of technological change on our values, therefore, depends on discovering the specific ways in which this process of accommodation occurs, on identifying the trends implicit in it, and on tracing its consequences for the value system of contemporary American society.

How Technology Leads to Value Change

In the article referred to above, Robin Williams defines values as "those *conceptions of desirable states of affairs* that are utilized in selective conduct as *criteria* for preference or choice or as *justifications* for proposed or actual behavior," and he distinguishes "some fifteen major value-belief clusterings that are salient in American culture, as follows: (1) activity and work; (2) achievement and success; (3) moral orientation; (4) humanitarianism; (5) efficiency and practicality; (6) science and secular rationality; (7) material comfort; (8) progress; (9) equality; (10) freedom; (11) democracy; (12) external conformity; (13) nationalism and patriotism; (14) individual personality; (15) racism and related group superiority." [16]

It seems clear that values in this sense have their origins in the patterns of choice behavior that are characteristic of any given society. What we mean when we say that a society is committed to certain values is that the people in that society will typically make judgments and choose to act in ways that reveal and reinforce those values. It seems equally clear that choice behavior is determined, or at least circumscribed, by the options available to choose from at the time the choice is made. We can choose to go to the country or to go to the moon, but we cannot at this time choose to go on living for 150 years, because that option is not now available to us.

Available choice options do change over time, of course. Thirty years ago we could not have chosen to go to the moon; 30 years from now we may succeed in extending the human life span to 150 years. When options are thus changed or expanded, it is to be expected that choice behavior will change, too, and changed choice behavior can in turn be expected, given appropriate time lags, to be conceptualized or "habitualized" into a changed set of values.

Technology has a direct impact on values by virtue of bringing about just such changes in our available options. By literally creating new opportunities for action, it offers individuals and society new options to choose from. Space technology—to stay with the most conspicuous example—makes it possible for the first time to go to the moon or to communicate by satellite, and it thereby adds those two new options to the spectrum of choices available to society.

By adding new options in this way, technology can lead to changes in values in the same way that the appearance of new dishes on the heretofore standard menu of one's favorite restaurant can lead to changes in one's taste and choices of food. Specifically, technology appears to lead to value change either by bringing some previously unattainable goal within the realm of choice, or by making some values easier to implement than in the past, that is, by changing the costs associated with realizing them.

These and related hypotheses are being examined in a number of studies that are under way as this is written.[17] For example, Dr. Irene Taviss of the Harvard University Program on Technology and Society is exploring the ways in which technological change affects the intrinsic sources of tension and potential change in value systems. When technology facilitates implementation of some social ideal and society fails to act on this new possibility, the conflict between principle and practice is sharpened and leads to new or aggravated tensions.

Thus, the economic affluence that industrial technology has helped to bring to American society makes possible fuller implementation of our traditional values of social and economic equality than was possible in the past. Unless or until it is acted upon, however, that very possibility gives rise to the tensions we associate with the rising expectations of the underprivileged and provokes both the activist response of the radical left and the modern hippie's rejection of society as hypocritical and unwilling to practice in reality what it preaches in its ideals.

Another example of the effects of technological change on values is related to our concept of democracy. The ideal we associate with the old New England town meeting is that each citizen should have a direct voice in political decisions. The ideal was never realized even in eighteenth-century New England, of course, yet it expresses a prominent characteristic of our political value system. Since realization of the ideal has not been possible in fact, we have elected representatives to serve our interests and vote our opinions.

Sophisticated computer technology now makes possible rapid and efficient collection and analysis of voter opinion, however, and could eventually provide for "instant voting" by the whole electorate on any issue presented to it via television a few hours before, for example by the President of the United States. It thus raises the possibility of instituting a genuine system of direct democracy.

On the face of it, this possibility promises implementation of the old New England ideal and tends to appeal to segments of the population that are advocating greater popular participation in the political process. On the other hand, it arouses the apprehension of those who see an important distinction between genuine democratic participation and quick opinions quickly delivered and who treasure the potentials for public education and sober political judgment implicit in our existing system of governmental checks and balances, political debates, legislative delays, and judicial reviews.

The technical possibility of instituting a system of direct democracy via computerized voting raises still another issue, namely, whether the majority of people are interested or willing to accept responsibility for expressing themselves directly on the many complex and technical questions with which modern government has to deal. Most people expect to be consulted and like to vote on major questions of policy, but they would probably find it unduly burdensome to be saddled with the responsibility of deciding the million detailed and technical questions that constitute the bulk of the work of government. In another area, some investigators have found that most patients, when confronted with the opportunity of participating in decisions affecting their own medical care, prefer to rely on the technical expertise of their physicians. There is a minority, no doubt, that would welcome a voice in the details of their own treatment and of government policy, but it seems doubtful that they could command general agreement on either question.

In other words, such potential technologies as computerized instant voting—and such actual ones as television news, for that matter—have the effect of raising political tensions and of challenging our society to clarify the meaning that it attaches to democracy. Do we construe democracy as the un-tutored will of an undifferentiated majority, as the result of transient coalitions of different interest groups or conflicting value commitments, as the considered judgment of the people's elected representatives, or as by and large the mixed system of government we actually have in the United States, minus the flaws in it that we would like to correct? By bringing us face to face with such questions, Dr. Taviss argues, technology has the effect of calling society's bluff and thereby preparing the ground for changes in its values.

In the case where technological change alters the relative costs of implementing different values, it impinges on inherent contradictions in our value system. To pursue the example explored immediately above, modern technology can enhance the values of participation that we associate with our democracy. But it can also enhance another American value—that of "secular rationality," No. 6 in Robin Williams' list—by facilitating the use of scientific and technical expertise in the process of political decision making. This can in turn lead to a further reduction of citizen participation in the democratic process. Technology thus has the effect of facing us with contradictions in our own value system and of calling for deliberate attention to their resolution. The problem is aggravated by the regrettable but undisputed fact that our social and moral sciences are as yet far from being equipped, conceptually or methodologically, to deal with such issues in a concrete and rigorous way.

The Value Implications of Economic Change

The changes discussed and illustrated in the preceding section represent the relatively direct effects of technology on values. In addition, value change often comes about through the mediation of some more general social or cultural change produced by technology. In such cases, the effect of technological change on values can be thought of as being indirect. Two examples of such indirectly induced value change are being explored respectively by Professor Nathan Rosenberg of Wisconsin University and by Professor Harvey Cox of the Harvard Divinity School. These are the value changes that result from the effects of technology on the economy and the value changes that result from the effects of technology on our religious belief systems.

In the discussion of the negative effects of technology at the end of Chapter I above, I alluded to the tension imposed on our individualistic values by the external benefits and costs of technological development. Professor Rosenberg is exploring the closely allied problem that is posed for an individualistic value system by the need for society to provide what economists call public goods and services.

Public goods and services differ from private consumer goods and services in that they are provided on an all-or-none basis and consumed in a joint way, so that more for one consumer does not mean less for another. For example, the clearing of a swamp or a flood-control project, once completed, benefits everyone in the vicinity. A meteorological forecast, once made, can be transmitted by word of mouth to additional users at no additional cost. Knowledge itself may be the prime example of a public good, since the research expenses needed to produce it are incurred only once, unlike consumer goods of which every additional unit adds to the cost of production.

As noted earlier, private profit expectation is an inadequate incentive for the production of public goods and services, because their benefit is "indiscriminate" and therefore not fully appropriable to the firm or individual that incurs the cost of producing them. Individuals are therefore motivated to dissimulate by understating their true preferences for such goods, in the hope of shifting their cost to others. This creates a "free-loader" problem, which skews the mechanism of the market. The market therefore provides no effective indication of the optimal amount of such public commodities from the point of view of society as a whole. If society got only as much public health care, flood control, or knowledge as individual profit calculations would generate, it would no doubt get less of all of them than it does now or than it expresses a desire for by collective political action.

This gap between collective preference and individual motivation imposes strains on a value system, such as ours, that is primarily individualistic rather than collective or "societal" in its orientation. That system arose out of a simpler, more rustic, and less affluent time than ours, when both benefits and costs were of a much more private sort than now. What public goods were needed, it was assumed, could be provided largely by private individual enterprise and action, given only a liberal constitution and a governmental guarantee of individual rights.

That assumption, and the individualistic value system it supported, are no longer fully adequate for our society, which industrial technology has made productive enough to be able to allocate significant resources to the purchase of public goods and services and in which modern transportation and communications as well as the sheer magnitude of technological effects lead to extensive ramifications of individual actions on other people and on the environment.

The response to this changed experience on the part of the public at large generally takes the form of increased government intervention in social and economic affairs to contain or guide these wider ramifications, as noted previously. The result is that the influence of values associated with the free rein of individual enterprise is put under strain. Society finds that it must strike a new balance between individual rights and the public interest and create the new social mechanisms needed to institutionalize it. This process will be accompanied, inevitably, by a change of orientation of the traditional American value system in the direction of legitimating the growing emphasis on collective, or society-wide, decision and action. To be sure, the tradition that ties freedom and liberty to a laissez-faire system of decision making remains very strong and the changes in social structures and cultural attitudes that can touch it at its foundations are still only on the horizon. But the trend seems clearly implicit in the imperatives of technological change.

Religion and Values

Much of the unease that our society's emphasis on technology seems to generate among some sectors of society can perhaps be explained in terms of the impact that technology has on religion.

The formulations and institutions of religion are not immune to the influences of technological change, for they, too, tend toward an accommodation to changes in the social milieu in which they function. One of the ways in which religion functions, however, is as an ultimate belief system that provides legitimation, that is, a "meaning" orientation, to moral and social values. This ultimate meaning orientation, according to Harvey Cox, is even more basic to human existence than the value orientation. When the magnitude or rapidity of social change threatens the credibility of that belief system, therefore, and when the changes are moreover seen as largely the results of technological change, the meanings of human existence that we hold most sacred seem to totter, and technology appears to be the villain.

Religious change thus provides another mediating mechanism through which technology affects our values. That conditions are ripe for religious change at the present time is evident in the spectrum of events from the "Death-of-God" movement among Protestant theologians to the questioning of the authority of the Pope in moral matters among segments of the Catholic clergy. Such developments amount to asking whether our established religious syntheses and symbol systems are adequate any longer to the religious needs of people who are living in a generally scientific and secular age that is changing so fundamentally as to strain traditional notions of eternity. If they are not, how are they likely to change? In the study alluded to above, Professor Cox is addressing himself to this problem, with specific attention to the influence of technology in channeling the direction of change.

He finds that the generation and the use of technology are so much a part of the style and self-image of present-day society that men begin to experience themselves, their power, and their relationships to nature and history in terms of open possibility, hope, action, and self-confidence. The symbolism of such traditional religious postures as subservience, fatefulness, destiny, and suprarational faith begins then to seem irrelevant to our actual experience. This symbolism loses credibility, so that its religious function is weakened.

When that happens, secular belief systems arise to compete for the allegiance of men: such political belief systems as communism, which reject divinity in favor of one or another alternative object of faith; or scientific ones, such as modern-day ethical culture movements; or such inexplicit, noninstitutionalized belief complexes as are characteristic of agnosticism or atheism.

Not only must religion somehow meet the challenge of such "loss of faith," however; it needs also to come to terms with the pluralism of explicitly religious belief systems that is characteristic of the modern world. No religion that hopes to be influential today can be formulated in terms that ignore the parallel existence and competitive appeal of alternative faiths. This is so, partly because we have learned a good deal about how different religions function in different societies and partly because widespread knowledge of other peoples makes social and religious diversity seem less unnatural and more acceptable than was the case in earlier times.

This pluralism of belief systems poses serious problems for the ultimate legitimation or "meaning" orientation of moral and social values that religion seeks to provide, because it demands a religious synthesis that can integrate the fact of variant perspectives into its own symbol system. Western religions have been notoriously incapable of performing this integrating function and have rather gone the route of schism and condemnation of variance as heresy. The institutions and formulations of historical Christianity in particular, which once provided the foundations of Western society, carry the added burden of centuries of conflict with scientific world views as these competed for ascendancy in the same society. This makes it especially difficult for traditional Christianity to accommodate to a living experience so infused by scientific knowledge and attitudes as ours, and it helps explain why its adequacy is coming under serious question at the present time.

Cox has noted three major traditions in the Judeo-Christian synthesis and has found them inconsistent in their perceptions of the future: an "apocalyptic" tradition foresees imminent catastrophe and induces a negative evaluation of the world; a "teleological" tradition sees the future as the certain unfolding of a fixed purpose inherent in the universe itself; a "prophetic" tradition, finally, sees the future as an open field of human hope and responsibility and as becoming what man will make of it.[18]

Technology, as I have noted, creates new possibilities for human choice and action, but leaves their disposition uncertain. What its effects will be and what ends it will serve are not inherent in the technology but depend on what man will do with technology. Technology thus makes possible a future of open-ended options that seems to accord well with the presuppositions of the prophetic tradition. That tradition may therefore provide us with the context necessary to a new religious synthesis that is both adequate to our time and continuous with what is most relevant in our religious history.

Arriving at such a new synthesis would of course require an effort at deliberate religious innovation for which Cox finds insufficient theological ground at the present time, despite the attempts represented by Vatican II and other recent ecumenical movements. Although it is recognized that religions have changed and developed in the past, conscious innovation in religion has generally been condemned and is not provided for by the relevant theologies. The main task that technological change poses for theology in the next decades, therefore, is that of deliberate religious innovation and symbol reformulation to take specific account of religious needs in a technological age.

What consequences would such changes in religion have for values? Too little is known as yet about the relationship of religion and values to allow anything like a comprehensive answer to all aspects of so broad and fundamental a question. Considering the principal importance of the symbolic dimension of religious experience, however, examination of the effects of technology on specifically expressive or aesthetic values—rather than on social or moral values, in the first instance— seems particularly promising. Cox therefore attempts the beginning of an answer to the general question in the context of the familiar complaint that, since technology is principally a means, that is, a tool, it enhances merely instrumental values at the expense of expressive or consummatory values that are considered to be somehow more "real."

The appropriate distinction is not between technological instrumental values and nontechnological expressive values, however, but among different expressive values that attach to different technologies. The horse-and-buggy was a technology too, after all, and it is not self-evident that its charms were different in kind or superior to the sense of power and adventure, and the spectacular views, that go with jet travel.

In fact, technological advance is in many instances a condition for the emergence of new creative or consummatory values. Improved sound boxes in the past and structural steel and motion photography in the present have made possible the artistry of Jascha Heifetz, Frank Lloyd Wright, and Charles Chaplin. These artists have opened up wholly new ranges of expressive possibility without, moreover, in any way inhibiting a concurrent renewal of interest in medieval instruments and primitive art.

One concludes that the new religious synthesis we seek would similarly forge new symbols expressive of technological reality. It could thus show the way in which society's values might be brought into better accord with contemporary experience. Indeed, if religious innovation can provide a meaning orientation broad enough to accommodate the idea that new technology can be creative of new values—or can serve to enhance or provide new content for old values—a long step will have been taken toward providing a religious belief system adequate to the realities and needs of a technological age. And it might be quite enough if religion performed this function only in the realm of expressive values. The challenge to education in general and to the humanistic disciplines in particular would then be clear: to accomplish a similar transformation of values in the social and moral realms.

CHAPTER III

Economic and Political Organization

The Enhancement of the Public Sphere

When technology brings about social changes (as described in Chapter I) that impinge on our values (in ways reviewed in Chapter II), it poses for society a number of problems that are ultimately political in nature.

The term "political" is used here in the broadest sense: It encompasses all of the decision making structures and procedures that have to do with the allocation and distribution of wealth and power in society. The political—or, better yet, the economic/political—organization of society thus includes more than the various bodies officially charged with public decision making; that is, it includes more than formal government. It also includes the market mechanism and other economic institutions, and it includes private decision making as well—by firms, or labor unions, or churches, or political parties, or professional or trade associations, or by individuals, for that matter—to the extent that such private decisions affect or have

63

implications for public decisions (i.e., for public policy).

The term "political" is thus a bridging concept; it bridges public and private organizations, formal and informal procedures, small and large groups. It is particularly important to attend to the organization of the entire body politic in this way when technological change contributes to a blurring of once clear distinctions between the public and private sectors of society and to changes in the roles of its principal institutions.

What is the import of big science and of powerful technologies for the political organization of modern society? More generally, what happens when the new rationality, the proliferation of expertise, and the growing social importance of knowledge run head on into political structures, processes, attitudes, values, and practices that developed in the hundred years between Andrew Jackson and World War II, that is, during a time when science and technology were not so big and powerful as they have become since and when the social role of knowledge and its institutions was secondary at best or virtually nonexistent at worst?

If we look for an answer in terms of a basic and long-term trend, I think we must conclude that a major effect of an active science and technology and of a commitment to knowledge as an instrument of social action is a progressive enhancement of the range and influence of the public sector of society in general, and of public decision making in particular. A number of considerations lead to this conclusion.

For one thing, the development and application

of technology seem increasingly to require large-scale and complex social concentrations, whether these be large cities, large corporations, big universities, or big government. In instances where technological advance appears to facilitate reduction of such first-order concentrations, it tends instead to enlarge the relevant *system* of social organization, that is, to lead to increased centralization. Thus, the physical dispersion made possible by transportation and communication technologies tends to enlarge the urban complex that must be governed as a unit.

A second characteristic of advanced technology is that its effects cover large distances, in both the geographical and the social senses of the term. Both its positive and negative effects are more extensive. For example, horse-powered transportation technology was limited in its speed and capacity, but its nuisance value was also limited, in most cases to the owner and to the local townspeople. The supersonic transport airplane would carry hundreds across long distances in minutes, but its noise and vibration damage would also be inflicted, willy-nilly, on everyone within the limits of a swath 3,000 miles long and several miles wide.

The concatenation of increased density, extended "distance," and multiplying population means that technological applications have increasingly wider ramifications, and that increasingly large concentrations of people and organizations become dependent on technological systems. A striking illustration of this was provided by the widespread effects of the power blackout in the northeastern part of the United States. The result, as noted in the discussion of technology assessment above, is that more and more decisions that could once be left to private decision makers because their effects were limited in impact and extent must now be taken in public ways, by society as a whole. Education, medicine, population policy, as well as the conduct of science and technology are additional examples of domains that were once largely private but are now increasingly coming into the public sphere.

In the eighteenth and early nineteenth centuries the getting and providing of education were entirely private affairs, whether engaged in by individuals or by private groups such as churches or philanthropic associations. Education began to move into the public realm, however, with the adoption of compulsory schooling and the support of schools by public funds. We see a further push in that direction today. As knowledge and technical skills become important requirements for entry into society and as the leaders of society (whether in industry, government, or any other social sector) must be better and better trained to cope with the complexities of the modern world, we see the proportion of public investment in education rising.

We also see educational standards increasingly set at the national rather than at the state or local level, and we see the control of educational policy and practice moving away from local publics—local boards—and toward national authorities. Not much of this is formal, of course, since there is the Constitution to contend with, but formal and informal regulation, voluntary raising of educational standards, and the structuring of incentives (such as by laying down requirements for Federal aid) are all pushing in that direction. When knowledge becomes increasingly important in society, education increasingly becomes a public concern and leaves the realm of the private.

The same trend is evident in the delivery of medical care, albeit much more recently in America than in most parts of Europe.[19] Once considered a privilege to be sold and paid for on an individual basis, medical care is increasingly regarded as a right that society is expected to finance and provide to its members. The delivery of medical care is thus becoming public in character, and the process receives added stimulus from the fact that modern medical technology makes the private, small shopkeeper practices of the past both socially and technically inefficient.

Other examples of this shift from private to public arise in connection with the determination of population size, which has traditionally depended on circumstances and private family decision, and scientific research and development, the course and magnitude of which used to depend entirely on private decisions by scientists about what to investigate and by industrial firms about what to develop and introduce into the market.

Population control is now on the verge of becoming a public concern, as a result of the dangers of overpopulation and the possibilities of avoiding them that are inherent in new birth-control technologies. In the case of research and development, what kinds of and how much science will be done in the country and what technologies—at least major and costly technologies—will be developed and applied are already largely matters of public decision.

The trend illustrated by such examples impinges on the American political tradition in a way that often makes it difficult for us to see what is happening. That tradition has produced two basic philosophies about the proper role of government in society. Depending on whether one has been a laissez-faire type-Republican or a socially-minded Democrat, government—in the popular phrase—was properly on tap or on top. Those were never pure types, of course, but they did define a spectrum in terms of which one could take a political stand.

Nowadays, however, this kind of typology is simply becoming irrelevant, precisely because of the new dominance of the public sphere I have been describing. To interpret this dominance as a victory of a Hamiltonian over a Jeffersonian concept of the proper role of government—as Senator Goldwater sought to do in his presidential campaign in 1964, for example—is to misunderstand and distort what is happening by trying to force it into old categories. For, as noted at the beginning of this chapter, the public or political sphere is much broader in scope and wider in extent than government in the formal sense.

What is actually happening is best described as a mixing-up of social institutions. Put more positively, what we are seeing is the forging of new partnerships between governmental and nongovernmental forms, as all institutions in the society become aware of the increasingly public character of the problems they face. In these new partnerships, government finds a new dimension, a new role, that we have not normally associated with it. No longer is government either the simple arbiter of conflicting interests between business, labor, farmers, or whatever, or the agent to whom all social action should be delegated. Instead, government—however haltingly as yet—is taking on the function of social pioneer and leader of a team; it seeks to identify opportunities over the horizon and problems before they are upon us and to marshal the forces, public and private, needed to deal with both. This helps explain the futuristic cast of many contemporary government programs. It also helps explain the intersectoral mechanisms that are being devised to enable cooperative efforts to deal with new developments as they emerge.

These changes are coming about in response to the growing need to make our social decisions deliberately and in public ways, rather than allowing them to "fall out," so to speak, of the impersonal interplay of innumerable private decisions. That need will continue to grow as technology, economic affluence, and increasing population combine to multiply both the opportunities and problems that society faces and to accelerate the changes with which it must come to terms. This means that allowing political change to come gradually and of its own accord may no longer be a viable strategy for contemporary society, as many of our youth are coming to insist. Instead, we face the problem of deliberately restructuring our political institutions and decision making mechanisms—including the system of economic decision making—to make them adequate to the enhanced social role of the public sphere. For example, the public goods problem alluded to earlier generates a need for such institutional innovation in the organization of the economy, as does the advent of modern computer-based information-handling techniques in the area of governmental decision making.

Private Firms and Public Goods

The public goods problem arises, it will be recalled, as a result of the shift in the composition of aggregate demand in the society at large. That is, the problem arises because the demand for public goods and services—such as education, health, transportation, slum clearance, and recreation facilities—increases relatively to the demand for private consumer goods and services. This shift in demand comes about in part because technological innovation creates new possibilities that we want to take advantage of as a society and in part because technology has enhanced our productivity so that we can afford to take advantage of many of them.

To be sure, we Americans choose to trade some of our increased productivity for more leisure, in the form of shorter hours, earlier retirement, or longer vacations. We also use some of it to buy more consumer goods and services: a second car, or a third television set, or steak twice a week instead of once. Even after that, however, we have enough national resources remaining to want to buy a number of public goods and services—clean water and air, moon shots, aid to less developed nations, and amelioration of domestic poverty are further examples—the value of which we increasingly appreciate and that we can buy only as a public by collective social decision.

I dealt briefly in the last chapter with the tensions created for our individualistic value system by this growing demand for public goods and services. The political face of the problem arises because the mechanism of the market, which is remarkably efficient in providing consumer goods, is, by general agreement, much less so in providing public goods. That is why governments have usually played a role in providing public goods, even though their performance too has often left a good deal to be desired. The shift in demand in favor of public goods, therefore, raises serious questions about the traditional roles of business and government in our society.

Robin Marris of the University of Cambridge has noted that, in the United States and other western industrialized countries, new technological developments have generally originated in and been applied through joint stock companies whose shares are widely traded on organized capital markets. Corporations thus play a dominant role in the development of new methods of production, of new products to satisfy consumer wants, and even of new wants. Most economists are inclined to accept the thesis first advanced by Joseph Schumpeter that such corporate activity is the key to the actual process of technological innovation in the economy. Marris himself has recently characterized this activity as a perceiving of latent consumer needs and a fostering and regulating of the rate at which these are converted into felt wants.[20]

There is no similar agreement about the implications of all this for social policy. J. K. Galbraith, for example, argues that the large modern corporation is motivated principally by the desire for growth, subject only to a level of profitability adequate to maintain management incomes and keep stockholders happy. He infers from this thesis that new-want development occurs at a higher rate than would be the case if corporations were motivated principally to maximize profits and he discerns a bias in favor of economic activities heavy in "technological content" (for example, new science-based consumer products) in contrast to activities requiring sophisticated social organization (for example, stimulating the economy of the urban ghetto). From this, Galbraith concludes that there is a bias in the economy generally in favor of development and satisfaction of private needs to the neglect of public needs and, therefore, a relatively slow rate of innovation in the public sector.[21]

The problem that these various findings and considerations point to can be restated in a manner that makes clear the issue of economic/political organization that is involved. Thus, corporations have proved to be highly efficient in exploring technological possibilities for new consumer products and services that can be marketed and sold for a profit. There are a number of reasons for this efficiency. The system of feedback and reward is speedy and highly visible; the corporation whose product does not sell quickly shifts its product line or goes bankrupt. It is this tension between the possibility of great success and quick failure, in fact, that creates the presumption that few technological possibilities for the private sector will go unexplored and that few of the wrong guesses will remain hidden for very long. Because of this kind of built-in efficiency, the corporate system has served us well—better than most, one is inclined to agree—when our greatest need as a society was feeding, clothing, and sheltering our population and raising our standard of living, that is, when our greatest need was for translating our technological progress into an abundance of private goods and services.

By contrast, there exist only relatively inadequate institutional mechanisms devoted to exploring technological possibilities for new socially desirable public goods and services. Traditionally, such mechanisms have included government agencies, corporations that serve public ends indirectly in the course of responding to a market or performing on government contract, mixed bodies such as port or river authorities, and combinations of public and private organizations.

As noted above, these institutional mechanisms tend to perform reasonably well in response to national crises or when the derivative personal benefit is clear, as in the provision of public health, for example. They have generally performed well also in the provision of the public counterparts of private goods, as in public roads for private automobiles. They have not been effective by and large, however, in guarding against the more general social costs of technology or in meeting needs— such as for adequate housing or environmental betterment—the satisfaction of which involves institutional rivalry and political conflict. Nor have they been aggressive in searching out technological possibilities for opportunities that society might well elect to choose if they were there to see. They have not been motivated to invent "new dishes" for the social menu.

A part of the reason for the latter failure, of course, is that such possibilities are difficult to see, because what we do not have and never have had is much more difficult to see or to miss than what we do have or what we may have had and lost. The full benefits of genuinely imaginative and creative uses of the mass media, for example, or the full potentials of public transportation technology cannot be appreciated until such time as they might be realized. Another reason is that public institutions do not have means equivalent to advertising and market research to inform the public about new possibilities and to test the potential demand for them. Finally, there is the difficulty of measuring the potential benefits of envisaged public goods and of deciding whether they are worth their costs. The discipline of the profit-and-loss statement is lacking, as are the clear feedback and the incentive of calculable rewards. Exactly analogous considerations, as we have seen, reduce the efficacy of these institutional mechanisms in containing the negative effects of technological change.

The need for institutional innovation that I cited above arises out of these inadequacies of present institutions. With the proper economic and political organization, we could derive greater benefits from our technology than we do. We might not then simply wait for them to fall out as incidental consequences of entrepreneurial activities but would be able to pursue them directly as a matter of deliberate public policy. That could stimulate development and application of technologies aimed primarily at producing social benefits and only secondarily at generating private profits. Such a reversal of our traditional priorities could help, not only to reduce pollution or rebuild our cities, but also to create new social opportunities in education, in health, or in cultural development.

Unfortunately, the knowledge and the ideas necessary to chart the necessary institutional innovation are largely lacking still, mainly because the recognition of the need is only very recent. In response to this lack, Dr. Marris is currently engaged in a study designed to formulate the relevant research questions and suggest some early answers.[22]

Among the questions this study is addressing are: (1) the effects of corporate goals and organization on the provision of socially desirable goods, (2) the costs to society of a policy aimed principally at stimulating economic growth rather than at a broader and more balanced social development, (3) the incommensurability of individual incentive and public will, (4) the desirable balance between individual and social welfare when the two are inconsistent with each other, (5) changes in the roles of government and industrial institutions in the political organization of American society, and (6) the consequences of those changes for the functions of advertising and competing forms of communication in the process of public education.

In particular, attention is being directed to

whether existing forms of company organization are adequate for marshaling technology to social purposes by responding to the demand for public goods and services, or whether new productive institutions will be required to serve that end. Other investigators are beginning to address one or another aspect of this problem too, of course, but it will be some time yet before these efforts begin to add up collectively to anything approximating a reasonable blueprint for the social innovation that we need.

Scientific Decision Making

A need for innovation in our political institutions and attitudes is implicit also in the new methods and tools of decision making that science and technology are making available to government.

One way in which this is coming about is through the proliferation of what might be called scientific or knowledge agencies in government itself: examples in the Federal government are the National Science Foundation, the National Aeronautics and Space Administration, and the Atomic Energy Commission (as specifically scientific or technological agencies) and the Departments of Health, Education, and Welfare, of Housing and Urban Development, and of Transportation (for what might be thought of as the more broadly knowledge-oriented agencies).

A second way in which science is affecting public decision making is by introduction of the professional scientist into the policy-making process, and in the development of greater technical sophistication on the part of the professional civil servant and government official. The broader face of this trend is the increasing use of experts of all sorts in the process of government, with the problems that this implies for the role of a less expert Congress and a less expert electorate in public decision making.

A third and increasingly important way in which governmental decision making is utilizing scientific techniques is by the introduction of computerized information-handling procedures and the adoption of some recently developed intellectual methodologies. In a recent study, Alan Westin of Columbia University identified the routes by which these two kinds of techniques are coming into government.[23] The first is through the introduction of computerized data banks, that is, automatic systems for storing, analyzing, and recalling information. A data bank is much better for this purpose than traditional filing systems and rows of human analysts, because it can handle much more information, is much faster, does not make mistakes, and does not forget.

Professor Westin identifies five principal types of data banks currently in use at all levels of government: statistical data banks for policy studies, mainly at the federal level; coordinating data banks for multiagency use; administrative data banks; statistical, regulatory, or subadministrative data banks for single-agency use below the chief executive level; and mixed public/private data banks in which information from both sources is pooled for a particular purpose, with restricted access.

The second development is the spread of the techniques of management science—operations research, systems analysis, and program planning and budgeting procedures—both in the Federal government and increasingly in state and municipal government. Recent instances of this have included contracts by the State of California to the aerospace industry for analyses of such problems as police protection and waste disposal and the continuing collaboration for analogous purposes of the City of New York and The RAND Corporation.

Professor Westin points out that reactions to the prospect of more and more decision making based on computerized data banks and scientific management techniques run the gamut of optimism to pessimism. Negative reactions take the form of rising political demands for greater popular participation in decision making, for more equality among all segments of the population, and for greater regard for the dignity of individuals. The increasing dependence of decision making on scientific and technological devices and techniques is seen as posing a threat to these goals, and pressures are generated in opposition to further "rationalization" of decision-making processes. These pressures have the paradoxical effect, however, not of deflecting the supporters of technological decision making from their course, but of spurring them on to renewed effort to save the society before it explodes under planlessness and inadequate administration.

The paradox goes further and helps to explain much of the social discontent that we are witnessing at the present time. The greater complexity and the more extensive ramifications that technology brings about in society tend to make social processes increasingly circuitous and indirect, as was suggested earlier above. The effects of actions are widespread and difficult to keep track of, so that experts and sophisticated techniques are increasingly needed to detect and analyze social events and to formulate policies adequate to the complexity of social issues.

The imperatives of modern decision making thus appear to require greater and greater dependence on the collection and analysis of data and on the use of technological devices and scientific techniques. Indeed, many observers would be likely to agree that there is an "increasing relegation of questions which used to be matters of political debate to professional cadres of technicians and experts which function almost independently of the democratic political process." [24] In recent times that trend has been most noticeable in the areas of economic policy and national security affairs.

These imperatives of modern decision making, however, run counter to that element of traditional democratic theory that places high value on direct participation in the political process and thus generates the kind of discontent mentioned above. If it turns out on more careful examination that direct participation is becoming less relevant to a society in which the connections between causes and effects are long and often hidden—which is an increasingly "indirect" society, in other words—elaboration of

a new democratic ethos and of new political institutions and procedures more adequate to the realities of a modern technological society will emerge
as perhaps the major intellectual and political challenge of our time.

At a minimum, it would seem that such new
political forms would have to embody a distinction
between the expression of preferences and the prediction of consequences. The opportunity to ex-
press preferences cannot, in a democracy, be properly denied to any segment of the population. The
more complex the society, in fact, and the more
technical the issues it has to deal with, the more
deliberate must be the effort to allow for the free
expression of preferences. The simple institution
of the quadrennial or biennial ballot on candidates
presumed to represent all the issues does not any
longer appear to be fully adequate to contemporary
need; our society may be outgrowing it as it is outgrowing the early value system that attended its
birth. A more differentiated system of electoral
consultation needs to be devised that will be responsive at short intervals when necessary, that can
distinguish among different kinds of issues, and
that can inform those whom it consults about
alternative options and their associated costs.

The need for a more refined system of electoral
consultation does not however imply that the
electorate at large, or any particular segment of it,
has a role to play in the technical process of government itself. In a populous, modern, industrialized, and knowledge-oriented society, that process
consists increasingly of adumbrating alternative
policy options and calculating their probable con-

sequences. It is clearly a job for experts and for all the sophisticated information-handling and management techniques that can be brought to bear on it. To be sure, the policy experts and decision makers must remain accountable to the electorate in the end. An adequate system of electoral consultation must therefore also provide effective mechanisms for monitoring and evaluating actual performance. But consultation and accountability remain distinct from the technical decision making process, and no amount or combination of them— that is, no amount of "participation" in the populist sense of the term—can substitute for the expertise and decision making technologies that modern government must use.

These considerations do no more than point to the general direction in which the needed restructuring of our political institutions must proceed. They do not begin to convey the difficulty of the intellectual and political problems that must be met before the actual restructuring can begin, nor do they evoke the strength of privilege and vested interest that will stand in the way. But institutional innovation here, as in the area of economic organization, is still in its early infancy. The hard work remains to be done.

Individual Rights and Public Responsibilities

What do technological change and the social and political changes that it brings with it mean for the life of the individual and the responsibilities of citizenship? It is not clear that their effects are all one way.

We are often told, by the pessimistic critics of technology alluded to in Chapter I, for example, that today's individual is alienated by the vast proliferation of technical expertise and complex bureaucracies, by a feeling of impotence in the face of "the machine," by a decline in personal privacy, and by the loss of an effective voice in the determination of public policy.

Although there is no way of proving the point, it is probably true that the social pressures placed on individuals today are more complicated than they were in earlier times. Increased geographical and occupational mobility can aggravate feelings of uncertainty in people by depriving them of the stabilities associated with living in one place or plying one trade over a lifetime. Also, the need to function in large organizations places difficult demands on individuals to conform or "adjust." This is often felt as degrading—by many of our youth, for example—because it is interpreted as a subordination of humanity to organizational efficiency.

It is also evident that individual privacy declines in a complex technological society, for a number of reasons. Many people voluntarily trade some of their privacy for benefits that they value more highly. For example, there is some abridgment of privacy entailed in having a telephone and being listed in the telephone book, in applying for social security benefits, or in enjoying the convenience of a credit card. There are also a number of ways in which individual privacy declines by involuntary invasion: either with our knowledge, as when we file an income tax return or register for selective service; or without our knowledge, as when we may be victims of wiretapping or any of the number of listening or visual surveillance devices that are available to anyone cheaply and by mail order. Such decline in individual privacy is probably an inevitable concomitant of the enhancement of the public sphere that was discussed earlier in this chapter.

Finally, it must certainly be conceded that the power, authority, influence, and scope of government are greater today than at any time in the history of the United States. The principal concerns of governments in the past have been to provide for national defense and to act as agents of social justice. Much more than that is demanded of modern government and much more is attempted by it. In today's highly industrialized mass societies, government takes on responsibility for education, for public health, for cultural development, for provision of housing, for the functioning of the economy, and for the support of scientific research and technological development. To the extent that government does so it encroaches on domains that were once exclusively private.

It might appear, therefore, that the prerogatives and political significance of the individual are in fact reduced if not submerged by big technology and complex social organization, as many contemporary social critics claim. But there is another, no less compelling side of the coin, which is cur-

rently being explored in a study by Edward Shils
of the University of Chicago.[25] For example, government may be more powerful and pervasive than
in the past, but it also appears to be more lacking
in confidence than ever before. As people become
more educated and self-confident they yield less to,
and demand more of, government. To the extent
that government is not fully responsive to that
demand it generates the kind of revolt against
authority that we see in all parts of the world and
thereby puts itself on the defensive. Where political
rulers once relied on charisma and appealed to
their traditional if not divine right to govern, they
now consult opinion polls, accept criticism from
press and public, and engage in dialogue with their
constituents. If we compensate for the large populations and structural complexity of modern mass
society, there is a strong presumption that the
average citizen today can "reach" and influence
his government more easily than could his counterparts in earlier societies.

There are also two sides to the privacy question.
While privacy may be declining in the various ways
indicated above, it also tends to decline in a sense
that most of us are likely to approve. The average
man in Victorian times, for example, probably
"enjoyed" much more privacy than today. No one
much cared what happened to him and he was
free to remain ignorant, starve, fall ill, and die in
complete privacy; that was "the golden age of
privacy," as Professor Shils puts it. Compulsory
universal education, social security legislation, and
public health measures—that is, the very idea of

a welfare state, which depends on availability of extensive and accurate information about the nation's population—are all antithetical to privacy in this sense, but it is the rare individual today who is loath to see that kind of privacy go.

It is not clear, finally, that technological and social complexity must inevitably lead to reducing the individual to "mass man" or "organization man." Economic productivity and modern means of communication allow the individual to aspire to more than he ever could before. Better and more easily available education not only provides him with skills and with the means to develop his individual potentialities, but also improves his self-image and his sense of value as a human being. This is probably the first time in two centuries that such high proportions of people have *felt* like individuals; few eighteenth-century English factory workers or nineteenth-century factory or farm workers in America, so far as we know, had the sense of individual worth that underlies the demands on society made by the average resident of the black urban ghetto today.

As Professor Shils points out, moreover, the scope of individual choice and action today are greater than in previous times, in the choice of consumer products, marital partner, occupation, place to live, objects of loyalty, and allegiance to religious, political, and other social groups. It should be emphasized that the computers and other information-processing and communication technologies that are said to submerge individuality

may, on the contrary, be essential to its maintenance. Professor Martin Shubik of Yale University
has made this point as follows:

> If we wish to preserve even modified demo-
> cratic values in a multi-billion-person society,
> then the computer, mass data processing, and
> communications are absolute necessities. . . .
> Using an analogy from the ballet, as the set
> becomes more complex and the dancers more
> numerous, the choreography required to main-
> tain a given level of co-ordination becomes far
> more refined and difficult. The computer and
> modern data processing provide the refine-
> ment—the means to treat individuals as in-
> dividuals rather than as parts of a large aggre-
> gate.[26]

Even the much maligned modern organization
may in fact "serve as a mediator or buffer between
the individual and the full raw impact of technological change." [27] The lack of concern for the
person that is cited as a failing of modern bureaucracy may thus have a silver lining after all. The
very impersonality of the big organization—like
the promise of anonymity that attracts so many
people to a big city—can help to protect individuality and personal freedom. There can be such a
thing as too much concern for the individual.

There are no doubt many times when the feeling
that individuality is being sacrificed to organizational efficiency is justified, but much of it may
also be another result of the application of old
values and old criteria to essentially new situations.

The fact is that technology and population growth do combine to reduce the scope and effectiveness of purely individual endeavor and to enhance the need for organized effort. But organized effort is not in itself demeaning, provided the participants in it understand their individual roles and learn to perform them well. I would doubt that the sense of achievement and quality of personal satisfaction of those who had a hand in the moon landing in 1969 were any less intense or different in kind from the feelings of Charles Lindbergh in 1927 or of Christopher Columbus and his band of adventurers in 1492. A good deal of the reason that many young people today suspect that they are, may be that we are failing to educate them properly about the complexity of the social roles they will be called upon to play and the opportunities that will be open to them when they are able to master those roles.

Recognition that the impact of modern technology on man as individual and as citizen has two faces, negative and positive, is consistent with the hypothesis about the double effect of technological change that was advanced in Chapter I. It also suggests that appreciation of that impact in detail may not be achieved in terms of old formulas, such as more or less privacy, more or less government, more or less freedom, or more or less individuality. To quote again from Martin Shubik:

The nature of government for a multi-billion-person world (and, eventually, planetary system) is neither quantitatively nor qualitatively the same as that required for an isolated New

England village. What freedoms do we intend
to preserve? Perhaps it would be more accurate to ask: What new concepts of freedom do
we intend to attach the old names to? [28]

Given the inadequacy of old formulas, it may be
that inquiry into the problem of human choice and
effective citizenship in a technological society is
more fruitfully conducted in terms of the kinds of
commitment that individuals living in such a society
are called upon to make. As at once individuals
and social beings, all of us eventually come to
some balance, appropriate to each of us, between
the relative degree of commitment we are prepared
to make to private and to public goals and values.
Each of us must achieve a symbiotic relationship
between our private and public selves, in other
words, by deciding about the degree to which we
will function as social beings and the degree to
which we will pursue private satisfactions.

The ancient texts we have suggest that the free
citizens of Periclean Athens were almost entirely
public beings in this respect; their major rewards
came from the public arena and even their enter-
tainments, such as sports and theater, were public
occasions. They literally lived in the market place.
They only slept in their houses and their purely
private pursuits and activities appear to have
counted for little. By contrast, certain segments of
today's youth seem by and large to turn their backs
on society and on public life, to reject their values
and responsibilities, and to devote themselves almost exclusively to the pursuit of private goals and

gratifications, whether physical, chemical/physiological, or psychological.

The Athenian and the hippie of the sixties thus define the spectrum along which each of us will choose where he will stand. Few of us choose either extreme but rather settle for some point in between. The enhancement of the public sphere that I have discussed, the trend away from an individualistic and toward a more collective value orientation, and the growing complexity of society—that is, the principal ways in which technological change affects society—would seem to call for a shift of the commitment point along that spectrum, away from the private in the direction of the public. How, how fast, and to what degree that shift will occur may determine how successful modern society will be in dealing with its technology.

Notes

1. The Harvard University Program on Technology and Society was established in 1964 by a grant from the International Business Machines Corporation to undertake an inquiry in depth into the effects of technological change on the economy, on public policies, and on the character of the society, as well as into the reciprocal effects of social change on the nature, dimension, and directions of scientific and technological developments.

2. Anthony G. Oettinger, with the collaboration of Sema Marks, *Run, Computer, Run: The Mythology of Educational Innovation,* in the series "Harvard Studies in Technology and Society" (Cambridge: Harvard University Press, 1969).

3. A preliminary report on this study will be found in Manfred Stanley, "The Technicist Projection: A Study of the Place of Social Theory in Moral Rhetoric," Jack D. Douglas, ed., *Freedom and Tyranny in a Technological Society* (New York: Random House, 1970).

4. Anne P. Carter, *Structural Change in the American Economy,* in the series "Harvard Studies in Technology and Society" (Cambridge: Harvard University Press, 1970).

5. *Report of the National Commission on Technology, Automation and Economic Progress,*

Volume I (Washington, D.C.: U.S. Government Printing Office, 1966), p. 9.

6. The findings of this study have been published in a number of articles and discussion papers originating in the Harvard University Program on Regional and Urban Economics. This program of studies was initiated by John R. Meyer, now at Yale University, and is currently headed by John F. Kain of the Harvard Department of Economics.

7. This study is awaiting publication by Harvard University Press in a volume on the social implications of biomedical science and technology that is planned for the series "Harvard Studies in Technology and Society."

8. Richard S. Rosenbloom and Robin Marris, eds., *Social Innovation in the City: New Enterprises for Community Development;* a collection of working papers published by the Harvard University Program on Technology and Society. (Cambridge: 1969; distributed by Harvard University Press.)

9. Emilio Q. Daddario, "Technology Assessment," U.S. House of Representatives, Committee on Science and Astronautics, Subcommittee on Science, Research, and Development, Ninetieth Congress. (Washington, D.C.: U.S. Government Printing Office, 1967.)

10. Louis H. Mayo, "The Technology Assessment Function," Part I. Program of Policy Studies in Science and Technology, The George Washington University. (Internal Reference Document, July 1968.)

11. Hearings on Senate Resolution 78, "To Establish a Select Senate Committee on Technology

and the Human Environment." U.S. Senate, Committee on Government Operations, Subcommittee on Intergovernmental Relations. Ninety-First Congress, March 4, 5, and 6, April 24, and May 7, 1969. (Washington, D.C.: U.S. Government Printing Office; printed for the use of the Committee on Government Operations.)

12. U.S. Department of Health, Education, and Welfare, *Toward a Social Report* (Washington, D.C.: U.S. Government Printing Office, 1969).

13. Daniel Bell, "The Idea of a Social Report," *The Public Interest,* No. 15, Spring 1969, p. 72.

14. For typical statements about our "knowledge" society, see Daniel Bell, *The End of Ideology* (New York: The Free Press, 1962), and "Notes on the Post-Industrial Society," I and II, *The Public Interest,* Nos. 6 and 7 (Winter and Spring 1967); and Robert E. Lane, "The Decline of Politics and Ideology in a Knowledgeable Society," *American Sociological Review,* Vol. 31, No. 5 (October 1966).

15. Robin Williams, "Individual and Group Values," *Annals of the American Academy of Political and Social Science,* Vol. 37 (May 1967), p. 30.

16. *Ibid.,* pp. 23 and 33.

17. The studies by Irene Taviss, Nathan Rosenberg, and Harvey Cox discussed in this chapter are parts of a project on technology and values being conducted by the Harvard Program on Technology and Society. An account of early findings appears in *Technology and Culture,* Vol. 10, No. 2 (April 1969), pp. 208–213.

18. Harvey Cox, "Tradition and the Future," I and

II, *Christianity and Crisis,* Vol. XXVII, Nos. 16 and 17 (October 2 and 16, 1967).

19. Reasons for the different rates of political development in the United States and Europe are discussed in Samuel P. Huntington, "Political Modernization: America vs. Europe," in Reinhard Bendix, ed., *State and Society: A Reader in Comparative Political Sociology* (Boston: Little, Brown, 1968).

20. Robin Marris, *The Economic Theory of "Managerial" Capitalism* (New York: The Free Press, 1964).

21. John Kenneth Galbraith, *The New Industrial State* (Boston: Houghton-Mifflin, 1967).

22. When this study has been completed, it is expected that its findings will be published by Harvard University Press in the series "Harvard Studies in Technology and Society."

23. Alan F. Westin, ed., *Information Technology in a Democracy,* in the series "Harvard Studies in Technology and Society" (Cambridge: Harvard University Press, forthcoming in 1970).

24. Harvey Brooks, "Scientific Concepts and Cultural Change," in G. Holton, ed., *Science and Culture* (Boston: Houghton-Mifflin, 1965), p. 71.

25. The study is being conducted for the Harvard University Program on Technology and Society.

26. Martin Shubik, "Information, Rationality, and Free Choice in a Future Democratic Society," *Daedalus,* Volume 96 (Summer 1967), p. 777.

27. Paul Lawrence and Jay Lorsch, *Organization*

and Environment: Managing Differentiation and Integration (Boston: Division of Research, Harvard Business School, 1967), p. 241.

28. Shubik, *op. cit.,* p. 776.

Annotated Bibliography

I. Social Change

Daniel Bell, "Notes on the Post-Industrial Society,"
Parts I and II, *The Public Interest,* 6 & 7
(Winter 1967 and Spring 1967), pp. 24–35
and 102–118.

The post-industrial society is characterized by a
service, rather than a manufacturing, economy
and by the centrality of theoretical knowledge.
The university replaces the business firm as the
central institution. Government decisions "will
replace the market as the arbiter of various social and economic choices" and there will be
more social planning. Despite the enhanced importance of technical decision making, the poli-
tician rather than the technocrat holds the
ultimate power; and "the irony is that the more
planning there is in a society, the more there
are open group conflicts."

Warren G. Bennis and Philip E. Slater, *The Temporary Society* (New York: Harper & Row,
1968).

In a time of widespread technological and social change, democracy becomes the most efficient form of social organization and science
emerges as a more suitable organizational

model than the military-bureaucratic model. Traditional bureaucracies are challenged by rapid change, growth in size, the complexity of modern technology and the internal diversity that results therefrom, and changes in managerial behavior. In their place, temporary systems emerge. The social and psychological consequences of temporary systems are explored.

Kenneth E. Boulding, *The Meaning of the 20th Century: The Great Transition* (New York: Harper & Row, 1964).

The first great transition of human history was that from precivilized to civilized society which began some five to ten thousand years ago; the second great transition is occurring now—it is the transition from the civilized to the post-civilized or technological society. The book examines the scientific and social scientific bases of this transition and the four traps that lie in it way: the war trap, the population trap, the technological trap (that is, the inability "to develop a genuinely stable high-level technology which is independent of exhaustible resources"), and the entropy trap (that is, the danger that man may "dissipate his energies in a vast ennui and boredom"). Accepting a "cautious and critical acceptance" of the transition, the author offers a strategy for handling it.

Peter F. Drucker, *The Age of Discontinuity* (New York: Harper & Row, 1969).

Four major discontinuities are examined: the development of "genuinely new technologies" and industries based upon them, the emergence of a world economy, the new sociopolitical reality of pluralistic, organized power concentrations, and the emergence of knowledge as the "central capital, the cost center, and the crucial resource of the economy." In each of the four areas, the changes are delineated and some policy recommendations are offered.

Jacques Ellul, *The Technological Society* (New York: Alfred A. Knopf, 1964).

The author takes a pessimistic view of the development of powerful and widespread modern technologies. He argues that technology has become an autonomous force that is self-perpetuating, beyond man's control, and destructive of human values.

Victor C. Ferkiss, *Technological Man: The Myth and the Reality* (New York: George Braziller, 1969).

The effects of current technological change on society are explored through an examination of "the direction in which mankind must move if it is going to be able to deal with the new challenges put to the social order by technological change." Technological man requires a philosophy that encompasses a new naturalism, holism, and immanentism. Social planning and awareness of the ecological consequences of policies are needed.

Victor Fuchs, "The First Service Economy," *The Public Interest,* 2 (Winter 1966), pp. 7–17.

This discussion of the growth of the service sector in the American economy examines the shift in employment from goods to services, the differences in the firm and in the nature of ownership between the goods-producing and the service sectors, and changes in the nature of work. The problems involved in measuring productivity and the lack of attention to the service sector by economists and statisticians are also discussed.

Michael Harrington, *The Accidental Century* (Baltimore: Penguin Books, 1965).

This book examines "the cultural and intellectual crisis confronting the United States and the rest of the Western world in the 20th century. The crisis has been brought about by the 'accidental revolution,' in which an unplanned social and creative technology has haphazardly reshaped our lives and put in doubt all our ideologies and beliefs." While the technological and social revolution of the twentieth century must continue, "it must cease to be accidental. It must become conscious, planned and democratic through the political intervention of man."

Harvard University Program on Technology and Society, "Implications of Biomedical Technology," *Research Review,* No. 1 (Fall 1968).

This is a review of the literature, with an introductory essay, summary materials, and lengthy abstracts of 55 items. The literature covered concerns policy for biomedical science, health and medical policy, and the social implications of genetic and behavior control, transplants and artifical organs, and human experimentation.

Harvard University Program on Technology and Society, "Technology and Work," *Research Review,* No. 2 (Winter 1969).

This is a review of the literature, with an introductory essay, summary materials, and lengthy abstracts of 46 items. The literature covered concerns the impacts of technology on the occupational distribution of the labor force and on work patterns and skills, and the public policy aspects of these changes.

Herman Kahn and Anthony J. Wiener, *The Year 2000: A Framework for Speculation on the Next Thirty-Three Years* (New York: Macmillan, 1967).

"By selecting extrapolations of current or emerging tendencies that grow continuously out of today's world," such as "the world-wide spread of a more or less secular humanism, the institutionalization of scientific and technological innovation, the expectation of continuous economic growth," the authors create a "surprise-free" projection of the year 2000. Consistent with this projection, they describe a "standard world" as well as some variations. Increased availability of goods, increased leisure, and technological changes in such areas as psychopharmacology will have important consequences for culture and styles of life. The society that will emerge—"affluent, humanistic, leisure-oriented, and partly alienated—might be quite stable."

Melvin Kranzberg and Carroll W. Pursell, Jr., eds., *Technology In Western Civilization* (New York: Oxford University Press, 1967), Volume I, *The Emergence of Modern Industrial Society, Earliest Times to 1900.* Volume II, *Technology in the Twentieth Century.*

Volume I details: The Emergence of Technology, Background of the Industrial Revolution, 1600–1750, The Industrial Revolution, 1750–1830, The Age of Steam and Iron, 1830–1880, and The Promise of Technological Fulfillment, 1880–1900. Volume II is addressed to: Rationalization and Its Consequences, Transportation, Materials and Structures, Energy Resources, Electronics and Communications, The Food Revolution, Land Use and Resources, Technology and the State, Technology in War, Scientific Research, Technology, and Automation, and Space, Culture, and Technology.

Marshall McLuhan, *Understanding Media: The Extensions of Man* (New York: Signet Books, 1964).

This book examines the social, cultural, and psychological consequences of electric technology. The author argues that "in the mechanical age now receding . . . slow movement insured that the reactions were delayed for considerable periods of time. Today the action and the reaction occur almost at the same time." Nevertheless, "we continue to think in the old, fragmented space and time patterns of the pre-electric age. . . . In the electric age, when our central nervous system is technologically extended to involve us is the whole of mankind . . . we neccesarily participate, in depth, in the consequences of our every action."

Emmanuel G. Mesthene, ed., *Technology and Social Change* (Indianapolis: Bobbs-Merrill, 1967).

This examination of the impacts of science and technology on modern society reprints Mesthene, Jacques Ellul, John Platt, Harvey Brooks, and Alvin Weinberg on the cultural implications of contemporary science and technology, Marshall McLuhan on electronics, Herbert Simon on computers, Joshua Lederberg on genetics and evolution, and Daniel Gershenson and Daniel Greenberg on the continuities in science.

Lewis Mumford, *The Myth of the Machine: Technics and Human Development* (New York: Harcourt, Brace & World, 1967).

Inventions in ritual, language, and social organization preceded the development of tools. This historical treatment of the development of tools begins with domestication and the early utilization of power. It argues that throughout history, man—and not his tools—has been dominant.

Lewis Mumford, *Technics and Civilization* (New York: Harcourt, Brace & World, 1963).

On the assumption that "to understand the dominating role played by technics in modern civilization, one must explore in detail the preliminary period of ideological and social preparation," the author examines the development of technology from the "first wave" of its spread around the tenth century. He argues that "before the new industrial processes could take hold on a great scale, a reorientation of wishes, habits, ideas, goals was necessary," and he seeks to answer the question of what are the necessary conditions if the machine is to "be directed toward a fuller use and accomplishment."

National Commission on Technology, Automation, and Economic Progress, *Technology and the American Economy* (Washington, D.C.: Government Printing Office, 1966).

This is the report of the President's Automation Commission, which examined the rates of technological change and their effect on economic growth and employment levels. It concluded that economic growth, rather than technological change per se, is the principal factor which determines level of employment, and offered policy recommendations for keeping the rate of growth high and for easing the problems of transition and unemployment which accompany technological change in specific sectors. Six Appendix Volumes detail: (1) The Outlook for Technological Change and Employment, (2) Employment Impact of Technological Change, (3) Adjusting to Change, (4) Educational Implications of Technological Change, (5) Applying Technology to Unmet Needs, (6) Impact of Technological Change.

Richard R. Nelson, Merton J. Peck, and Edward D. Kalachek, *Technology, Economic Growth, and Public Policy* (Washington, D.C.: The Brookings Institution, 1967).

"This book explores the relations among research, development, innovation, and economic growth; considers the manner in which the economy adapts to technical change and the problems encountered in the processes of adaptation; and recommends several policy changes designed to encourage technological change consistent with other public policy objectives."

Everett M. Rogers, *Diffusion of Innovations* (New York: The Free Press, 1962).

Through a comprehensive review of the literature the author emerges with some generalizations concerning the process whereby innovations are adopted and diffused. He examines the characteristics of the innovations and the innovators and the role of cultural norms in affecting the dissemination of new techniques and ideas.

Donald A. Schon, *Technology and Change* (New York: Dell, 1967).

This examination of the processes of technological innovation in industry and their social consequences points up the nonrational components of invention and suggests some necessary innovations in social norms. The norms and objectives of the process of discovery itself should become the basis for a social ethic to govern the process of change.

Irene Taviss, ed., *The Computer Impact* (Englewood Cliffs, N.J.: Prentice-Hall, 1970).

This reader examines the implications of computer technology for the economy, the polity, and the culture. The selections include discussions of computer application in industry, management, and decision making and white collar work; uses in government and the law, and attendant questions of privacy invasion; and impacts on the schools, branches of learning, and popular philosophy and attitudes.

"Toward the Year 2000: Work in Progress," *Daedalus*, 96 (Summer 1967).

The entire issue is devoted to the work of the Commission on The Year 2000 of the American Academy of Arts and Sciences. Daniel Bell, the chairman of the commission, provides an introduction and has selected excerpts from the transcripts of the working sessions of the Commission. Twenty-two essays on specific problem areas are included.

Charles R. Walker, ed., *Modern Technology and Civilization* (New York: McGraw-Hill, 1962).

This reader is divided into four parts. The first

offers some selections on the social history of technology; the second considers "the problems and promise of the machine age," especially the effects of new productive technologies on workers and organizations; the third is devoted to technology in non-Western countries; and the fourth examines the effects of technology on values and the psychology of man.

II. Values

Kenneth E. Boulding, "Dare We Take the Social Sciences Seriously?" *American Behavioral Scientist*, 10 (June 1967), pp. 12–16.

"Science is corrosive of all values which are based exclusively on simpler epistemological processes." The ethic of science and the ethic of other subcultures often conflict. But such incompatibilities can be sustained and may even be creative.

Kenneth E. Boulding, "The Interplay of Technology and Values: The Emerging Superculture," in Kurt Baier and Nicholas Rescher, eds., *Values and The Future: The Impact of Technological Change on American Values* (New York: The Free Press, 1969), pp. 336–350.

Both technologies (ways of doing things) and values (choice and preference processes) are "created and transmitted by a common learning process" and constantly interact. Both are changed in the course of transmission from one generation to the next. As a result of their complex interaction today, two cultural systems are emerging: the worldwide superculture of airports, throughways, and skyscrapers and the traditional cultures of various national, religious, ethnic, and linguistic groups. The two are interdependent and must learn to adapt to each other.

R. A. Buchanan, "The Religious Implications of Industrialization and Social Change," *The Technologist,* 2 (1965), pp. 245–255.

Science and technology have eroded traditional modes of thinking and traditional authorities; life has become demythologized. The churches must address themselves to the new problems of individual identity and the right use of leisure time, rather than bemoaning the depersonalization of an industrial society. Churches should treat their members as responsible and encourage maturity rather than submission to church authority.

Harvey G. Cox, "Tradition and the Future," I and II, *Christianity and Crisis,* 27 (October 2 and 16, 1967), pp. 218–220 and 227–231.

We have inherited from the Judeo-Christian tradition three different and contradictory perceptions of the future: the apocalyptic, teleological and prophetic. The apocalyptic tradition denies goals and rational action; teleology "obscures the fact that history is radically open"; and the prophetic tradition sees the future as open and man as having moral responsibility for it. "Only a recovery of the prophetic perspective will supply the ethos required for the political ethic required today."

Walter Firey, "Conditions for the Realization of

Values Remote in Time," in Edward Tiryakian, ed., *Sociological Theory, Values and Sociological Change* (New York: Harper & Row, 1963), pp. 147–159.

It is difficult to institutionalize "future-referring values" because the individual has a shorter time span and frame of reference than the society. If such values are to be implemented, individuals must find them both socially expedient and psychologically satisfying.

Thomas F. Green, *Work, Leisure, and the American Schools* (New York: Random House, 1968), Chapter 5, "Work and the Quest for Potency," pp. 115–146.

The current crisis of values is a crisis in the way in which values are experienced. The traditional moral rhetoric is no longer relevant. While our values still hinge on individualistic assumptions and morality is lodged in the individual, the moral agent today is the public agent. Effectiveness depends upon corporate and social action; and institutional means for holding people accountable must be developed.

Garrett Hardin, "The Tragedy of the Commons," *Science,* 162 (December 13, 1968), pp. 1243–1248.

Individualism and laissez-faire must be abandoned. What was moral a hundred years ago is no longer feasible today, because of population density and the ramifications of individual acts on the larger society. In the densely populated society of today, "mutual coercion mutually agreed upon" is required, and "freedom is the recognition of necessity."

Harvard University Program on Technology and

Society, "Technology and Values," *Research Review*, No. 3 (Spring 1969).

This is a review of the literature, with an introductory essay, summary materials, and lengthy abstracts of 44 items. The literature covered concerns the interaction of technology and values (generally and in the contemporary situation) and value problems in a technological society (including economic, political, and religious values, and the problem of social planning).

Robert Heilbroner, "Do Machines Make History?" *Technology and Culture*, 8 (July 1967), pp. 335–345.

A position of "soft determinism" is taken with respect to the question of whether machines make history. The argument is that while technology has a significant effect on social structure and values because it determines the composition of the labor force and the organization of work, technological advance is nevertheless itself responsive to social direction and must be compatible with existing social conditions. While up to now technology has been more determinative of social conditions than vice-versa, in the future man may assume more deliberate control over his technology.

Fred Charles Iklé, "Can Social Predictions Be Evaluated," *Daedalus*, 96 (Summer 1967), pp. 733–762.

If we are to make "guiding predictions," we must be able to "evaluate the desirability of predicted aspects of alternative futures." But since this desirability will be determined by our future preferences, we have to predict our values before predicting the future, and we must be able to shift back and forth between evaluation and prediction. Social predictions are also plagued by the fact that the social response to them often renders them either self-fulfilling or self-defeating.

Carl Kaysen, "The Business Corporation as a Creator of Values," in Sidney Hook, ed., *Human Values and Economic Policy: A Symposium* (New York: New York University Press, 1967), pp. 209–223.

The business corporation, through the market system, plays an important role in the process of value creation. In an affluent society in which the challenge of producing consumer goods has been met, more attention should be paid to the production of such public goods as education and health. The corporation should be more concerned with the values it is creating, and institutions of public, professional criticism should be developed. Some tempering of competition and of the reliance on pure market forces is needed.

Clyde Kluckhohn, "The Scientific Study of Values and Contemporary Civilization," *Zygon,* 1 (September 1966), pp. 230–243.

Values should be made congruent with what is known about nature and human nature. If scientific investigation can lead to knowledge of the consequences of various possible courses of action, man's values will be affected. Science can explore the relationships between instrumental and ultimate values in specific cultures and the cross-cultural aspects of values.

Melvin Kranzberg, "Technology and Human Values," *The Virginia Quarterly Review,* 40 (Autumn 1964), pp. 578–592.

The humanists are taken to task for having become "intellectual Luddites" and for failing to assume their responsibility of interpreting nature and society to man. Technology is a means that has helped to foster many democratic and humanistic values.

C. B. Macpherson, "Democratic Theory: Ontology and Technology," in David Spitz, ed., *Political Theory and Social Change* (New York: Atherton Press, 1967), pp. 203–220.

Western democratic theory has been based on two internally inconsistent assumptions. The first—associated with capitalism and the market —sees man as an infinite desirer and consumer of utilities. The second—which provided the justification for liberal democracy—views man as an exerter of his uniquely human capacities and asserts the equal right of every individual to make the most of himself. The two conflict because, if each man is allowed to consume in accordance with his infinite desires, some men will accumulate more than others and thereby gain power over others, so that the right of every man to make the most of himself will not be realized. The conflict can be resolved because the technological revolution allows us to abandon the market conception of man, since we are no longer faced with the problem of "enlist[ing] men's energies in the material productive process." To retain the values of liberal democracy, the market view of man must be abandoned.

Bruce Mazlish, "The Fourth Discontinuity," *Tech-*

nology and Culture, 8 (January 1967), pp. 1–15.

The failure to recognize that there is no discontinuity between man and the machine lies at the root of our current distrust of technology. We must transcend this discontinuity to allow for the harmonious acceptance of an industrialized world. To turn away from machines is to turn away from our own humanity.

Emmanuel G. Mesthene, "How Technology Will Shape the Future," *Science,* 161 (July 12, 1968), pp. 135–143.

Technology brings about both social and value change by creating new physical possibilities and thus altering the mix of social choices. By providing the means for bringing previously unattainable ideals within the realm of choice and by altering the relative ease with which different values can be implemented, new technology may produce a restructuring of the hierarchy of values. In an age of pervasive change more attention must be paid to process than to structure, and the emphasis should shift "from values to valuing."

Lewis Mumford, *Technics and Civilization* (Harcourt, Brace, World & Co., 1934), Chapter 7, "Assimilation of the Machine," pp. 321–363.

Man cannot simply impose his will upon machines. Machines have fostered "the technique of cooperative thought and action" as well as a new logic and esthetic. The machine process has created a new human value represented by "the concept of a neutral world . . . indifferent to [man's] activities." It is incumbent upon man first to absorb the lessons of the mechanical

realm—objectivity, neutrality, impersonality—
and then to go beyond them to "the more pro-
foundly human."

Hasan Ozbekhan, "The Triumph of Technology:
'Can' Implies 'Ought,'" System Development
Corporation, #SP-2830 (June 1967), 17 pp.

Planning runs counter to the basic Western traditions of laissez-faire equilibrium. As a result,
we have come to planning reluctantly and have
tended to see it as deterministic. A new theory
of planning is needed that will be open, rather
than deterministic, and that will deal with the
question of goals and values. There must be
normative, strategic, and operational planning
in order "to reveal what ought to be done, what
can be done, and what actually will be done."

Gunter W. Remmling, *Road to Suspicion* (New
York: Appleton-Century-Crofts, 1967), "Epi-
logue," pp. 199–210.

The modern mentality lacks any dominant values. There is multidimensionality and fragmentation, "warring universes of discourse." Value-
oriented images have become a poor guide to
behavior because they are easily corroded by
the rapid spread of scientific information.

Nicholas Rescher, "What is Value Change? A
Framework For Research," in Baier & Rescher,
eds., *Values and The Future: The Impact of
Technological Change on American Values*
(New York: Free Press, 1969), pp. 68–109.

Future economic and technological changes
will make some values easier and some more
difficult to realize. As a result some values will
be "upgraded" and others will be "downgraded." The results of a cost-benefit approach

to this question suggest that mankind-oriented values, intellectual virtues, rationality, group acceptance, social welfare, social accountability, order, public service, and aesthetic values will be upgraded, while nation-oriented values, economic and personal independence, self-reliance, self-advancement, individualism, economic security, and optimism will be downgraded.

Philip Rieff, *The Triumph of the Therapeutic* (New York: Harper & Row, 1966).

Contemporary culture is characterized by a release from all cultural and moral demands. There is an extreme "plasticity and absorptive capacity." Science, which is "supra-cultural" and "non-moral," embodies the moral revolution of our time. It cannot supply a creed to the nonscientists.

Lynn White, Jr., *Machina Ex Deo* (Cambridge: The M.I.T. Press, 1968), Chapter 2, "The Changing Canons of Our Culture," pp. 11–31.

The very "canons of our culture" have been changing. From viewing Western civilization as the ultimate in civilization, we are moving to a "canon of the globe"; the primacy of logic and language have been giving way to the primacy of symbols and that of reason to the unconscious. From conceptions of a hierarchy of values, we have been moving toward a spectrum of values in which all are equal.

Robin M. Williams, Jr., "Individual and Group Values," *The Annals of the American Academy of Political and Social Science,* 371 (May 1967), pp. 20–37.

Information concerning the distribution of values in American society would be of assistance in understanding current social problems and in formulating national policies. In a knowledge-oriented society traditional values and norms are increasingly subjected to questioning and reformulation.

III. Economic and Political Organization

Francis M. Bator, *The Question of Government Spending: Public Needs and Private Wants* (New York: Collier Books, 1962).

This book examines "some of the much-debated implications of government spending," while recognizing that "the question of what level of government spending is tolerable . . . cannot, in the end, be resolved by recourse to fact or logic. The answer will be, must be, sensitive to values: values as to the allocation and distribution of material resources, and, perhaps more important, as to the modes of social and political organization." To provide some clearer understanding of the consequences of feasible alternatives, a test is offered for "the two-pronged claim that any substitution of political for private market choice through increased public spending will, in general, 'worsen' the allocation of resources and that, in fact, government spending in the postwar period has been the cause of gross misallocation."

Raymond A. Bauer, ed., *Social Indicators* (Cambridge: The M.I.T. Press, 1966).

This collection of essays attempts to explain

the need for social indicators analogous to the economic indicators currently in use. They assess the current state of the art and offer proposals for implementing "social systems accounting."

James D. Carroll, "Science and the City: The Question of Authority," *Science* 163 (February 28, 1969), pp. 902–911.

The development of managerial and social technologies, as exemplified in the city planning work of the Department of Housing and Urban Development, is faced with obstacles from labor, political groups, and congressional wariness to urban research and development. While long-term planning is necessary, the question of who defines "the good" is raised. "The most difficult question is whether existing or new knowledge and technological processes will be of much use if the objectives for which these should be used cannot be defined clearly through the American political system."

"The Computer and Invasion of Privacy," Hearings Before a Subcommittee of the Committee on Government Operations, U.S. House of Representatives, Eighty-Ninth Congress (Washington, D.C.: U.S. Government Printing Office, 1966).

These hearings were held in June 1966 in response to the bill proposed by the Bureau of the Budget to establish a National Data Center.

Emilio Q. Daddario, "Technology Assessment," U.S. House of Representatives, Committee on Science and Astronautics, Subcommitte on Science, Research, and Development, Ninetieth Congress (Washington, D.C.: U.S. Government Printing Office, 1967).

This is a proposal that the Congress establish a Technology Assessment Board so that "policy determination in applied science and technology" will become "anticipatory and adaptive rather than reactionary and symptomatic." Technology assessment is "a form of policy research which . . . systematically appraises the nature, significance, status, and merit of a technological program."

John Kenneth Galbraith, *The New Industrial State* (Boston: Houghton Mifflin, 1967).

The economic and political consequences of technological change are based upon the "imperatives of technology" toward planning, specialization, and organization. This book examines the effects of these imperatives on the modern business firm, the "technostructure" at its head, the market, and the relationships between business and government. Among the book's major theses are that the market has declined as a guiding influence in economic life as planning has taken hold and that organized intelligence has replaced ownership as the source of power in the modern organization.

Robert L. Heilbroner, *The Limits of American Capitalism* (New York: Harper & Row, 1966).

While the current system of American capitalism is likely to be maintained through this century and well into the next, technological change is producing social changes that might render this system inadequate in the future. The new professional, scientific, and technological elites that are arising may attain sufficient power to substitute other goals and values for those of the business system. Also, business may not be able to handle the problems of material abundance and labor displacement which are brought by technological change.

Olaf Helmer, *Social Technology* (New York: Basic Books, 1966).

Based on the assumption that "the gap between physical technology and sociopolitical progress" can be narrowed, this book examines the new techniques of forecasting scientific and social innovation and the systematic use of expertise.

Paul T. Heyne, *Private Keepers of the Public Interest* (New York: McGraw-Hill, 1968).

This book examines the concept of the social responsibilities of business and finds it inadequate because it fails to deal satisfactorily with the economic problem of resource allocation and with the relationship between power and justice. It also obscures the real problems confronting the businessman who wants to act ethically.

Alfred J. Kahn, *Theory and Practice of Social Planning* (New York: Russell Sage Foundation, 1969).

This book attempts to examine how planning "can, should, and sometimes does take place." It offers "a rational model for the social planning process," while maintaining an awareness of the importance of the nonrational aspects of interpersonal, intergroup, and interorganizational processes. The major phases in the planning process—definition of the task, policy-formulation, programming, evaluation, and feed-

back—are analyzed, as well as the problem of
how planners cope with issues of fact and value.

Carl Kaysen, "Data Banks and Dossiers," *The Public Interest,* 7 (Spring 1967), pp. 52–60.

"While the fears raised by critics" of the proposal to establish a National Data Center "have
real content, the problems are neither entirely
novel, nor beyond the range of control by adap-
tations of present governmental mechanisms. . . .
The data center would supply to all users, in-
side and outside the government, frequency
distributions, summaries, analyses, but never
data on individuals. . . ." Technical safeguards
can be established against abuses. The data
bank is needed to remedy the defects in the
Federal statistical system, which is currently
"too decentralized to function effectively and
efficiently."

Robert E. Lane, "The Decline of Politics and Ideology in a Knowledgeable Society," *American
Sociological Review,* 31 (October 1966), pp.
649–662.

Our political norms and values have been
changing because "the political domain is
shrinking and the knowledge domain is growing, in terms of criteria for decisions, kinds of
counsel sought, evidence adduced, and nature
of the 'rationality' employed." There has also
been a reduction in ideological and dogmatic
thinking. But knowledge is a factor that ultimately produces disequilibrium, since it "creates
a pressure for policy change with a force all
its own."

Fritz Machlup, *The Production and Distribution of*

Knowledge in the United States (Princeton, N.J.: Princeton University Press, 1962).

This books presents an analysis of the "knowledge industries." The allocation of human, fiscal, and natural resources in knowledge production are assessed in the areas of education, research and development, communications media, information machines, and information services. The output of each field is evaluated in terms of its productivity, growth, and contribution to the body of knowledge.

Donald N. Michael, "Speculations on the Relation of the Computer to Individual Freedom and the Right to Privacy," *The George Washington Law Review,* 33 (October 1964), pp. 270–286.

Computer technology makes it easier to collect meaningful data about persons from a diversity of sources and makes it difficult for individuals to correct or provide interpretations for the records of their own past histories. But centralization and computerization of private information may also decrease the risk of illegitimate leaks of information. In the future, we may develop "a new measure of privacy: that part of one's life which is defined as unimportant (or especially important) simply because the computers cannot deal with it."

Arthur Selwyn Miller, "The Rise of the Techno-Corporate State in America," *Bulletin of the Atomic Scientists,* 25 (January 1969), pp. 14–19.

Constitutional values are threatened by technology and the concomitant rise of the supercorporation and the Positive State. The supercorporation is a center of economic power and consequently an instrument of government, though under the guise of private collectivism. The Positive State has engendered economic planning which blurs the line between private and public. As the technocorporate state has evolved, it has left old problems unresolved and generated new ones. These include: inflation, a tendency toward autarchy, government by elites, and problems of legitimacy and accountability.

James Schlesinger, "Systems Analysis and the Political Process," *Journal of Law and Economics,* 11 (October 1968), pp. 281–298.

The rationality or ethic of American political life often run counter to the rationality of such techniques as systems analysis. In systems analysis, a long-run goal is chosen, and careful calculation and measurements are then made to determine what the programs established to implement this goal are actually achieving. The rationality of politics, on the other hand, often dictates a focus on more short-run goals, the "appearance" of effort, and a thin spreading of resources over many problems in order to gain support from diverse groups. Moreover, because of the tendency among government agencies to withhold information that puts them in an unfavorable light, there are problems involved in the initial collection of information.

Martin Shubik, "Information, Rationality and Free Choice in a Future Democratic Society," *Daedalus,* 96 (Summer 1967), pp. 771–778.

While the assumptions of a rational, utilitarian man, the "invisible hand," and the democratic vote have been the basis of a free-enterprise democracy, these concepts are being called into question by changes in society and in knowledge. In a highly complex and information-rich society, "the limitations of the individual become more marked relative to society as a whole." And government in a multibillion-person world cannot be "the same as that required for an isolated New England village. What freedoms do we intend to preserve? Perhaps it would be more accurate to ask: What new concepts of freedom do we intend to attach the old names to?"

U.S. Department of Health, Education and Welfare, *Toward a Social Report* (Washington, D.C.: U.S. Government Printing Office, 1969).

This is a report presented to President Johnson in response to his request that yardsticks be developed for measuring the social state of the nation. Using a variety of social indicators, the essays examine and assess the state of affairs in health, social mobility and opportunity, the physical environment, income and poverty, public order and safety, learning, science and art, and participation and alienation.

Harold R. Walt, "The Four Aerospace Contracts: A Review of the California Experience," in Appendix Volume V of *Technology and the American Economy,* Report of the National Commission on Technology, Automation, and Economic Progress (Washington, D.C.: U.S. Government Printing Office, 1966), pp. 43–73.

This study reviews the conception, results, and implications of four studies that applied systems engineering techniques to public problems. The areas of application were waste management, crime and delinquency, information systems, and transportation.

Alan F. Westin, ed., *Information Technology in a Democracy* (Cambridge: Harvard University Press, 1970).

This reader explores the effects of information systems, systems analysis, and new techniques of decision making on the organization and nature of American government. The editor discusses the major types of information systems and decision making techniques, how and where they are being established, and what their likely impacts will be. The selections include descriptions of such systems and techniques; commentaries on their utility, value, and appropriateness; and some analyses of the social system into which they are being introduced.

Alan F. Westin, *Privacy and Freedom* (New York: Atheneum, 1967).

Recent advances in electronic, optical, acoustic, and other sensing devices have raised some threats to individual privacy. This book examines the capabilities of the new technology, the nature and meaning of privacy and the society's need to acquire information and to control individual behavior. Policy suggestions are offered to devise a system that could "identify and permit beneficial uses of the new technology and yet, at the same time, preclude those uses that most men would deem intolerable."

Aaron Wildavsky, "The Political Economy of Efficiency: Cost-Benefit Analysis, System Analysis, and Program Budgeting," *Public Adminis-*

tration Review, 26 (December 1966), pp. 292–310.

"Because the cost-benefit formula does not always jibe with political realities—that is, it omits political costs and benefits—we can expect it to be twisted out of shape from time to time. Yet cost-benefit analysis may still be important in getting rid of the worst projects. Avoiding the worst where you can't get the best is no small accomplishment."

Harold L. Wilensky, *Organizational Intelligence* (New York: Basic Books, 1967).

This book examines the uses of technical and ideolological "intelligence" in industry and government. "The quality of intelligence and its flow from the source to the user" are important because they affect "the performance of policy." The economic, political, and cultural contexts of intelligence are explored, and a military analogy is applied to a wide range of cases in order "to provide a perspective for the study of problems in the organization of the intelligence function common to all complex social systems."

Michael Wollan, "Controlling the Potential Hazards of Government-Sponsored Technology," *The George Washington Law Review,* 36 (July 1968), pp. 1105–1137.

Using weather modification, the supersonic transport, and fluoridation as case studies, the author concludes that "the federal government's vested interests in the continuation of its technological programs limit its ability to provide adequate technology assessment." As alternatives, it is suggested that "private groups, operating outside the framework of government, might be able to persuade it to place more emphasis on reduction and elimination of potential hazards of technology," and that within government, Congress might "establish a Technological Hazards Board, authorized by statute to appear before Congress and the agencies solely as a lobbyist for reduction and control of potential risks to the public and the environment."

Index